2008 Distinguished Instructor Short Course
Distinguished Instructor Series, No. 11

sponsored by the
Society of Exploration Geophysicists
European Association of Geoscientists & Engineers

# Reservoir Geophysics:
## Applications

presented by
**William L. Abriel**

**Society of Exploration Geophysicists**
*The international society of applied geophysics*

**EAGE**
EUROPEAN ASSOCIATION OF GEOSCIENTISTS & ENGINEERS

ISBN 978-1-56080-086-6 (Series)
ISBN 978-1-56080-146-7 (Volume)

Copyright © 2008
Society of Exploration Geophysicists
P.O. Box 702740
Tulsa, OK USA 74170-2740

Published 2008
Reprinted 2013

All rights reserved. No part of this book may be reproduced,
stored in a retrieval system, or transcribed in any form or by any means,
electronic or mechanical, including photocopying and recording, without
prior written permission of the publisher.

Printed in the U.S.A.

Library of Congress Cataloging-in-Publication Data

Distinguished Instructor Short Course (2008 : Tulsa, Okla.)

  Reservoir geophysics : applications : 2008 Distinguished Instructor Short Course / presented by William L. Abriel ; sponsored by the Society of Exploration Geophysicists [and] European Association of Geoscientists & Engineers.

     p. cm. -- (Distinguished instructor series ; no. 11)

  Includes bibliographical references and index.

  ISBN 1-56080-146-8 (volume : alk. paper) -- ISBN 1-56080-086-0 (series : alk. paper)

  1. Prospecting--Geophysical methods--Economic aspects--Congresses. 2. Oil fields--Valuation--Congresses. 3. Petroleum--Geology--Congresses. 4. Oil reservoir engineering--Decision making--Congresses. I. Abriel, William L., 1953- II. Society of Exploration Geophysicists. III. European Association of Geoscientists and Engineers. IV. Title.

TN271.P4D565 2008

622'.1828--dc22

2008010120

# Dedication

It would not be possible for this book to have been written without the inspiration I receive from the persons closest to me in life. For more than 50 years, my father, William Earl Abriel, has guided me as an unsurpassed role model. I discovered a few new things about life, but the majority of timeless lessons was given to me by him and continuously reinforced through the decades. It is not possible for me to be appreciative enough.

My immediate family (children Annie Laurie, William, James, and my beautiful and loving wife, Evangeline Abriel) has been the foundation of my life for more than 25 years, and their support has made this present work possible. Vangie has worked to encourage me in everything I do, giving me the confidence to progress when success seems so distant. She is an endless wellspring of golden value and an inspiration beyond measure.

# SEG and EAGE

*wish to thank the following
for their generous contributions*

**SEG Foundation**

**Chevron**

# About the Author

**William L. Abriel**, internal geophysical consultant at Chevron Energy Technology Company, San Ramon, California, began his work in the industry with Chevron in New Orleans in 1978. His technical interests are mainly in application of new technology to active projects, including acquisition, processing, interpretation, and integration. He has been the geophysical lead for Chevron projects in many oil and gas basins around the world, concentrating on North America, China, Australia, South America, and Russia. He has produced publications and has given presentations explaining and advancing geophysical technology.

Abriel has participated in technical committees for many SEG meetings on international, national, research, and multidisciplinary scales. He has published in most industry journals annually and has been a member of the editorial board of THE LEADING EDGE and an associate editor of GEOPHYSICS. He also has served on numerous SEG committees, including Development, Membership, Research, Global Affairs, and Distinguished Lecture. He was the SEG Spring Distinguished Lecturer in 2004 and is a founding and current board member of the SEG Advanced Modeling Corporation (SEAM). Abriel was named a life member of SEG in 2007.

He received a B.S. in geosciences and an M.S. in geophysics, both from Pennsylvania State University, where he was a founding member of the SEG student section and earned four varsity letters in lacrosse. He still keeps a hand in coaching lacrosse. He and his wife, Vangie, an attorney who teaches at Santa Clara University School of Law in California, have three children.

Dear DISC Participant:

It is a great pleasure to welcome you to the eleventh annual SEG/EAGE Distinguished Instructor Short Course (DISC), "Reservoir Geophysics: Applications," by William L. Abriel. SEG, EAGE, and your local society are proud to provide this premier course in geophysics education.

With rapidly changing technologies, geoscientists around the world have an increasing need to acquire expert technical knowledge. SEG's Professional Development program — the SEG/EAGE DISC, the SEG Distinguished Lecturer, the SEG/AAPG Distinguished Lecturer, the SEG Regional Lecturers, and other live and online short courses — aids in the promotion of technologies that will have a significant impact on geophysics and geoscience. Likewise, EAGE supports geoscience education through its EAGE Education Tours, EAGE Education Days, Distinguished Lecturers, short courses, and Learning Geoscience e-learning initiative.

Previous DISC programs were:

- "Time-Lapse Seismic in Reservoir Management," by Ian Jack, in 1998
- "The Seismic Velocity Model as an Interpretation Asset," by Phil Schultz, in 1999
- "Shear Waves from Acquisition to Interpretation," by Robert Garotta, in 2000
- "Seismic Amplitude Interpretation," by Fred J. Hilterman, in 2001
- "Understanding Seismic Anisotropy in Exploration and Exploitation," by Leon Thomsen, in 2002
- "Geostatistics for Seismic Data Integration in Earth Models," by Olivier Dubrule, in 2003
- "Petroleum Systems of Deepwater Settings," by Paul Weimer, in 2004
- "Insights and Methods for 4D Reservoir Monitoring and Characterization," by Rodney Calvert, in 2005
- "Seismic Attribute Mapping of Structure and Stratigraphy," by Kurt Marfurt, in 2006
- "Concepts and Applications in 3D Seismic Imaging," by Biondo Biondi, in 2007

Education in geophysics and geoscience is one of the top priorities for both SEG and EAGE. The DISC affords important opportunities for local geophysical organizations to provide first-rate geophysical education at modest cost. The program is truly a cooperative effort of many people dedicated to the promotion and advancement of geophysics. Your participation is key to the continued success of this annually renewed program.

We are honored to have William Abriel as instructor for our 2008 DISC program. This is a great opportunity to learn from one of our profession's recognized experts in the application of geophysical technologies for value optimization throughout the life cycle of a reservoir. We encourage you to take full advantage of this opportunity to broaden your perspectives through participation in the 2008 DISC.

Sincerely,

Fred Aminzadeh
SEG President

Finn Roar Aamodt
EAGE President

# Table of Contents

| | |
|---|---|
| Dedication | iii |
| Acknowledgments | viii |
| Chapter 1  Introduction | 1 |
| Chapter 2  Field Discovery and Delineation | 13 |
| Chapter 3  Delineation Problem | 29 |
| Chapter 4  Development | 41 |
| Chapter 5  Production | 53 |
| Chapter 6  Development Problem | 65 |
| Chapter 7  Heavy Oil | 81 |
| Chapter 8  Reservoir Geophysics in Carbonates | 91 |
| Appendix A  Salinas Problem | 107 |
| Appendix B  Cobra Problem | 115 |
| References | 121 |
| Index | 125 |

# Acknowledgments

This book is the result of a significant effort from a substantial number of active participants. Without their energetic, thoughtful, and involved contributions, this work would not exist.

Substantial additions to the technical material and improvements in communication of this textbook were made through active authoring and editing by the following, listed alphabetically: Chevron Energy Technology Company; Chevron International Exploration and Production, ONS asset team; Chevron North America Exploration and Production, Gulf of Mexico, Bay Marchand asset team; Chevron North America Exploration and Production, Gulf of Mexico, deepwater asset team; Alan Fuqua; William Haworth; George Hildebrandt; Thomas Hudson; Abigail Hymel; William Kempner; Thomas MacKinnon; Craig Mann; Bernard Regel; James Rickett; John Robinson; Ata Sagnak; and Ken Yeats.

The dedicated professionals of SEG are responsible for the quality and timeliness of the SEG/EAGE Distinguished Instructor Short Course and its accompanying publication each year — a feat which cannot be overestimated. I am thankful to the following for their work in editing, reviewing, logistics, and persistence in directing the publication process: Rowena Mills, Jennifer Cobb, Cecilia Martin, Ted Bakamjian, and Peter Pangman of SEG and editor Gary F. Stewart.

It is a special pleasure to thank Bach Dao of Chevron Energy Technology Company for his graphics support. I would not have gotten far without his guidance and his highly productive efforts.

Finally, I am indebted to the management of Chevron for interest, encouragement, and funding of the DISC program in the spirit of technical advancement for the geophysical community. Thanks to Sarah Saltzer, Aimee Edwards, William Hallager, Stephen Smith, Kenneth Nelson, and Mark Koelmel for their support.

— William L. Abriel
December 2007

# Chapter 1   Introduction

**Purpose**

The purpose of this SEG/EAGE Distinguished Instructor Short Course is to demonstrate how and why geophysics adds value in reservoir management. The intent is to show the strength and application of key geophysical tools, focusing on seismic data. Many excellent courses or references are available for seismic data, demonstrating how to acquire, process, or perform the mechanics of interpretion, so those topics are not covered here.

What is covered in this text is the relationship between the business opportunity of the petroleum operator and the application of seismic data to provide valuable subsurface information. The business frame is supplied by understanding the needs of the operator and how he will use the geophysically derived information. This is covered by discussing the operational phases of the reservoir (delineation, development…) in separate chapters and identifying what specific information is derived from geophysical (seismic) data that is useful in economic decisions.

Theoretical concepts of geophysics are not the focus of this work, despite their high importance. Excellent references exist for learning the underlying physics and sensitivities of most geophysical tools. Rather, this text attempts to teach by application, using stable and well-understood principles applied to interesting projects from around the world and in many geologic environments.

This textbook is organized with an introduction of terms and concepts, then a description of the reservoir delineation process, then a student problem in delineation. Next, the processes of development and production are described, followed by the second student problem. Finally, specialty applications are described for using geophysics in heavy-oil and carbonate reservoir projects.

**Definitions**

Start with only three definitions to introduce the material.

1) *Reservoir*: Subsurface body of rock with sufficient porosity and permeability to store and transmit commercial quantities of oil, gas, condensate, and mixtures of these substances.

2) *Geophysics*: The study of the earth by quantitative physical methods, especially by seismic reflection and refraction, gravity, magnetic, electrical, electromagnetic, and radioactivity methods (Sheriff).

3) *Reservoir geophysics*: The use of geophysical methods to assist in delineating or describing a reservoir, or monitoring the changes in a reservoir as it is produced. (Sheriff).

## The origins of the discovered hydrocarbon reservoir

With rare exceptions, hydrocarbons produced in the energy industry are generated from the decay and subsequent chemical alteration of organic materials. Oil and gas are generated from the breakdown over time of large quantities of very small animals and plants. During generation and migration, the hydrocarbons and waters expelled from compacting rocks move, usually upward, from higher-pressured regimes at depth to lower-pressured zones. Consequently, most hydrocarbons are not trapped in the subsurface. Usually, 80% to 95% or more are broken down by bacteria at depth or expelled at the surface, leaving the remaining amount trapped in reservoirs normally made of sandstone or carbonate. Those rocks can hold hydrocarbons in place when overlying rocks such as shale or anhydrite with significantly less permeability act as seals. Faults, unconformities, and lateral stratigraphic changes also act as seals.

## Geophysics

When the word *geophysics* is used in the oil and gas business, it almost always refers to reflection seismology, which is the focus of this course. Sound waves generated at the earth's surface propagate through a complicated geologic overburden. The energy reflected back to the surface is recorded and processed by highly sophisticated techniques and is interpreted by experts who make predictions of value to the oil and gas business (Figure 1). In this course, we will concentrate on how analyses of 3D seismic data are used to answer critical questions about subsurface uncertainty for oil and gas extraction.

As stated, this course will not cover how best to acquire seismic data or how to process data optimally. Those important topics deserve the detailed explanation they receive in other courses and textbooks. The very interesting and often valuable geophysical tools that supplement reflection seismic data are not covered in this course either. They include potential-field measurements of gravity, magnetics, electromagnetics, refraction seismic, and passive seismic recording. Those specialized tools should not be overlooked when addressing the potential of geophysical applications to reservoir management.

**Figure 1.** Seismic reflection data are obtained by recording the returns of sonic impulses. For marine acquisition, a ship uses air guns as a source and tows several recording streamers.

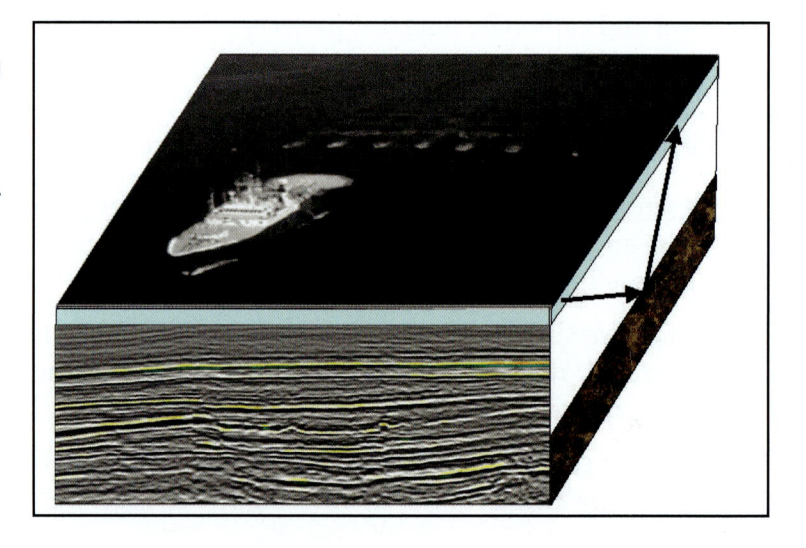

*Introduction*

During the past 60 years, seismic technology has evolved dramatically. At first, it was a tool to find large and shallow structures with seismic refraction methods. Then reflection data evolved to the point that structural geology could be mapped. When digital recording and processing emerged, seismic data began to provide stratigraphic information and even direct hydrocarbon detection. Initially, data were acquired with sources and receivers deployed in straight lines, which resulted in a 2D image of the subsurface. However, 2D data suffer from sideswipe and are a poor approximation compared with recording and imaging the subsurface in three dimensions.

When 3D seismic technology emerged, it revolutionized the industry because the ability to "see" the geology of the subsurface took a quantum leap (Figure 2). This significantly reduced the number of exploratory and developmental dry holes, and geophysicists provided project geologists with their first real look at widespread 3D stratigraphy.

However, 3D structure and stratigraphy were not the only things geophysics revealed. Geophysicists also began to provide reservoir engineers with time-lapse seismic surveys that detected fluid movement and pressure changes during production. Thus, seismic data evolved from primarily an exploration tool into one employed throughout the value chain of the oil and gas business, emerging as one of the leading impact technologies in the industry (Figure 3).

## Oil and gas operations

Because this course relates geophysics to the needs of oil and gas operators, a review of the activity for a reservoir project is useful. Starting with discovery, the reservoir management requires both delineation and development before production begins.

Oil and gas projects necessitate obtaining a history of investments that requires critical information about the subsurface (Table 1; Figure 4). The project begins at exploration discovery, when the geologic uncertainty of the project is greatest. At that point, very limited subsurface information is available to estimate the resource size and limits, and nearly all that information is derived from seismic data.

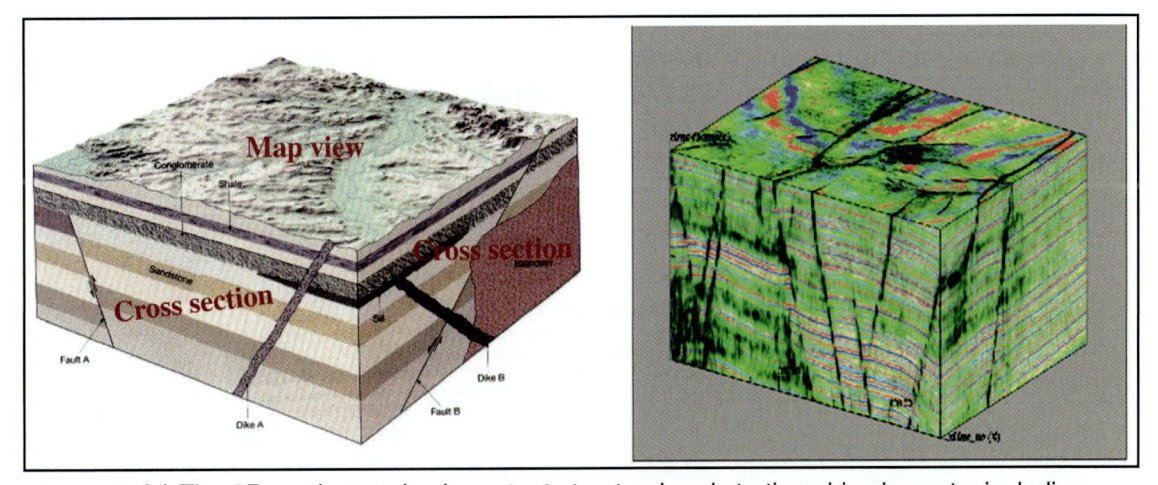

**Figure 2.** (a) The 3D earth contains important structural and stratigraphic elements, including the locations of trapped oil and gas reservoirs. (b) 3D seismic data can measure the acoustic response of the earth and reveal many of those elements.

From discovery, the project must proceed through the delineation phase. Stable estimates of the total hydrocarbon volume and deliverability are needed so the right size and type of facilities can be designed for extraction. That means moving into a delineation program that requires the drilling of wells to determine reservoir limits and total size. Delineation also must provide information about the restriction of flow resulting from faulting or lateral stratigraphic variations. As we will see, seismic data contribute significantly in that effort.

After delineation, if the resource is deemed economic, the project then enters the development stage, which includes installation of wells and production facilities. That is the stage with the highest investment level. Development wells must be drilled in the right places to efficiently drain the reservoir. Such wells are spotted very carefully to maximize production at minimum cost, and unexpected subsurface results during development drilling are not welcome. Again, seismic data are a substantial contributor to the process.

After development, the project goes into primary production, reaching peak flow rates until the reservoir energy (normally pressure) declines, production rates slow, and larger volumes of water flow into the producing wells. Under the right circumstances, reservoir energy can be enhanced through injection of gas or water in selected wells, and the project moves into secondary production support.

**Figure 3.** Interpretations of seismic data are now so important to oil and gas operators that they rank among the top impact technologies (data from Lehman Brothers, 2006).

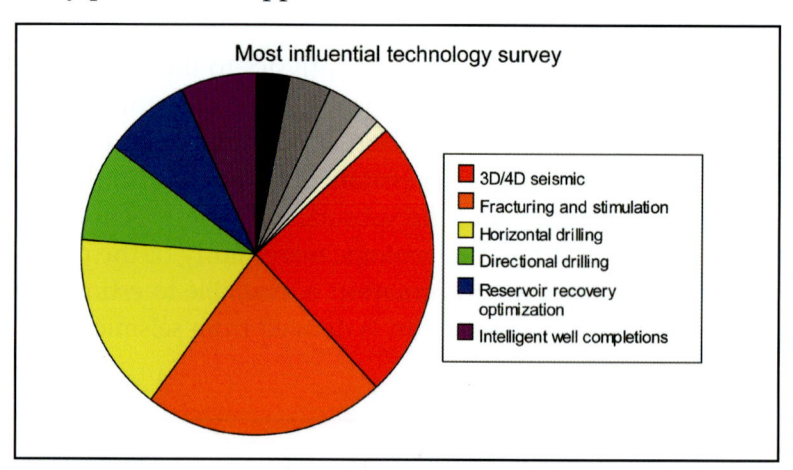

**Table 1.** Oil and gas operational phases and the information needed to make the investment.

| Project phase | Critical subsurface information |
| --- | --- |
| 1) Delineation | Total hydrocarbon volume<br>Areal limits of petroleum reservoir<br>Deliverability |
| 2) Development | Compartmentalization<br>Exact locations of development wells |
| 3) Production | Hydrocarbon saturation and pressure changes<br>Flow restrictions and channeling |

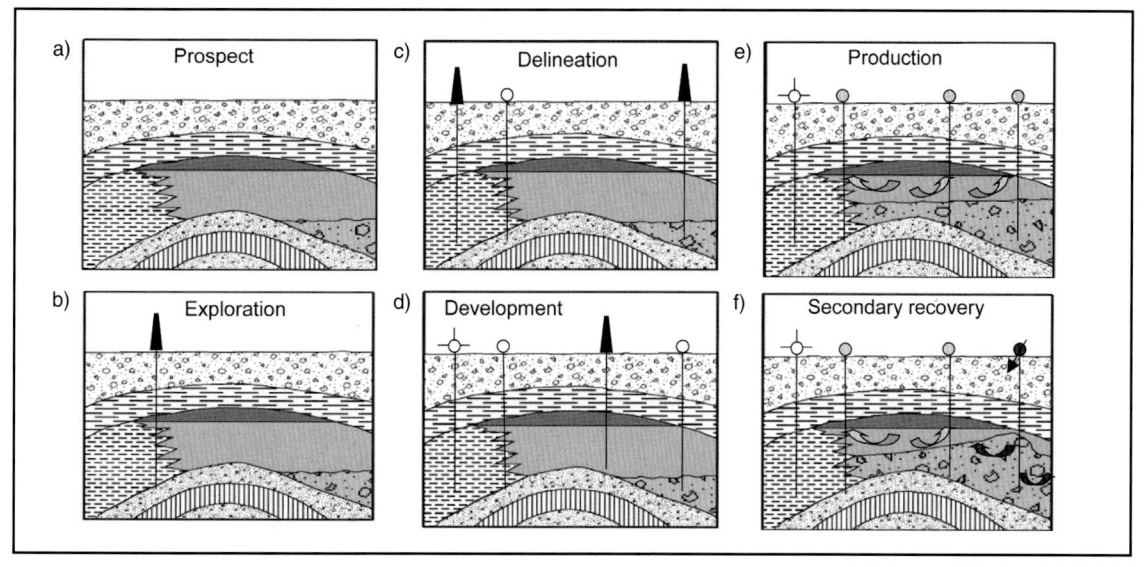

**Figure 4.** Management of the oil and gas field. (a) An exploratory prospect is primarily a geologic interpretation from seismic data. (b) Exploration is successful if a sustained flow of hydrocarbons is tested. (c) *Delineation* means drilling the optimal number of wells to find reservoir limits. (d) Development requires drilling the right number of wells to produce hydrocarbons. (e) Production extracts hydrocarbons, reduces pressure, and alters other fluid levels. (f) A secondary-recovery program can be employed with injection of water or gas to push more hydrocarbons to producers.

## How will geophysics help the oil and gas operator?

This course is an attempt to describe the role of geophysics in discovered reservoirs, as seen by oil and gas operators and their technical staffs. We will concentrate not on exploration but on the stages of delineation through production, focusing on what the oil and gas operator needs at each stage, what geophysics can do to help, and how to describe the uncertainties of geophysical predictions.

The focus will be on seismic data, principally 3D seismic data, the primary tool for understanding reservoir presence and "seeing" the properties between and beyond wells. Earth acoustic response is what is recorded and used in seismic analysis. Although it is an incomplete measure of the properties we want to learn about the reservoir, it is rich in information and can be understood in many ways.

## What reservoir properties do we want to predict?

From discovery to production, oil and gas operators work to recover hydrocarbons from far below the surface that they can neither see nor touch. To model and predict reservoir response, the operator needs to know critical reservoir characteristics. Characteristics before production are static properties, and those that affect fluid flow directly during production are dynamic properties.

Static properties include (1) fluid phase (oil and gas percent), (2) areal extent of the trap, (3) depth, (4) thickness, (5) compartmentalization, (6) reservoir net to gross, and

(7) porosity. Dynamic properties include (1) well deliverability, (2) reservoir connectivity, (3) permeability, (4) pressure changes, (5) phase changes, and (6) reservoir compaction.

Although some of those properties can be measured directly and accurately at the well, estimates away from the wellbore often become highly uncertain, and it is necessary to make predictions between and beyond wells. Often, that is accomplished in the context of a working geologic model through understanding of the structural style, stratigraphic setting, and fluid history.

## The role of seismic data in revealing earth properties

How does geophysics fit into the activity of reservoir prediction? Despite the advancement of well logging and downhole tools, seismic data are the only measurement of properties in the subsurface at a considerable distance from the wellbore. Thus, the data can be used to derive or modify the geologic and petroleum-engineering models used to manage the reservoir by providing a limited but highly useful measure of acoustic rock response. Dollar for dollar, obtaining information from seismic data is almost always considerably less costly than obtaining reservoir information from drilling more wells. However, seismic does not compete with the local extensive information that direct borehole data can provide, and it must merge with well data to be truly effective.

Seismic data are only the measure of earth response to wave propagation, but they yield a surprisingly great amount of information (Figure 5). In looking at seismic volumes of the earth, one sees how continuous reflections are shaped into large-scale forms interrupted by faults and unconformities, which provide structural information on the reservoir trap. Reflection patterns guide the interpretation of stratigraphic information, even to determining variations in reservoir depositional facies. Measured changes in reflection strength (amplitude) respond to lithology, porosity, and even fluid type (gas, oil, water).

Seismic data in two and three dimensions are interpreted in volumes to produce structure maps in time domain, which must be converted to depth by the earth velocity normally provided by seismic data away from wells. When the seismic horizon of the reservoir is tracked, attributes such as reflection strength from the reservoir are analyzed

**Figure 5.** A seismic cross section shows the elements of geology most used for subsurface reservoir interpretation. Structural information is derived from mapping continuous reflectors and faults. Stratigraphic knowledge is obtained from the patterns of reflectors caused by facies changes. Rock and fluid properties can be estimated from reflection strength.

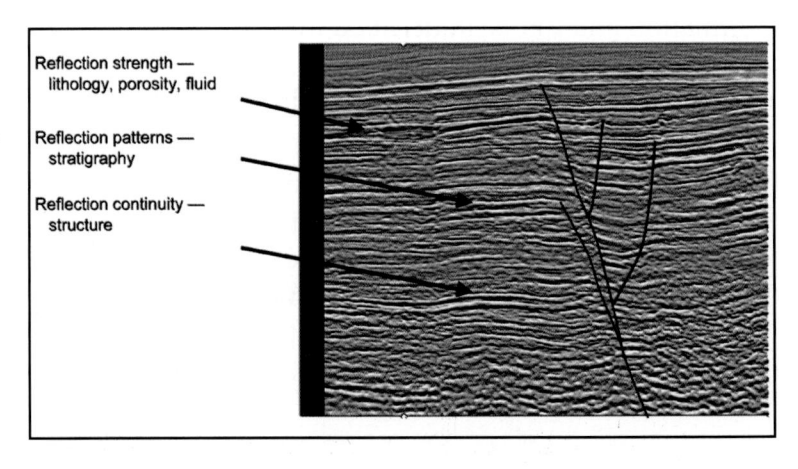

*Introduction*

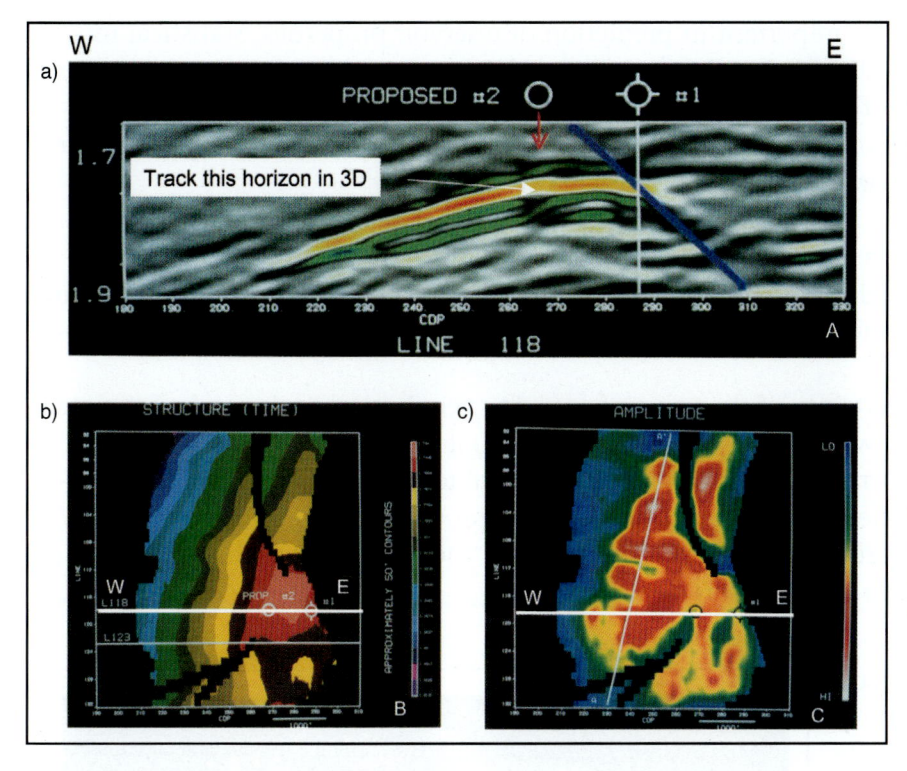

**Figure 6.** When (a) a seismic horizon is tracked in three dimensions, the time of the event is used to generate (b) a structure map. The amplitude of the tracked horizon, when seen in map form, is (c) the horizon slice. This shows horizontal differences in rock and fluid properties, with high-amplitude red representing the best porosity (Abriel and Wright, 2004).

to understand lateral differences in stratigraphy and reservoir fluids (Figure 6). Amplitude variations also are analyzed and are sorted by the angle at which they impinge on the reservoir (amplitude variation with offset [AVO]), and they can provide insights into lithology, porosity, fluid, and pressure.

Finally, analysis of the velocity of the rocks used to image seismic data provides additional information on lithology and pressure. However, it is important to remember that seismic data measure only the response of the rocks to the passage of sound waves. They are not a direct measure of those important reservoir parameters. Because of that, we must remember to provide uncertainty estimates along with the predictions of the reservoir that geophysics provides.

## Uncertainty tools used in geophysical analysis

Several tools are used to measure and/or communicate subsurface uncertainty. Visualization can be a very effective tool, especially for communication. Not all seismic data show the subsurface with the same clarity everywhere for a variety of reasons, including geologic complications of the near surface or scattering from complex structures. Visualization is a rapid and effective way of showing where subsurface prediction from seismic data can be strong or poor. Workstations and volume rendering of 3D data are used to communicate concepts of subsurface information of seismic data and are equally effective in communicating the variation of certainty of those concepts.

Along with visualization, it is common to illustrate uncertainty with models of wave propagation in the subsurface. For example, visualization can be an important tool to show why some subsurface areas are not illuminated well and why estimates of reservoir characteristics there are more uncertain (Figure 7).

Uncertainty is also important in prediction of reservoir properties. Statistical models of reservoir properties between wells can be used if the geologic concepts are appropriate and a solid database of spatial property changes can be drawn from well-studied examples elsewhere. Seismic data then can provide a further guide to constraining the uncertainties of reservoir parameters.

Uncertainties are important to take into account in the construction of reservoir models. An example is the calculation of reservoir volume (Figure 8). It is possible to make numerous models of the reservoir, each of which is equally likely. Thus the oil and gas operator can plan for managing a range of possible outcomes in delineation or production (Figure 9).

Finally, the uncertainty of the value of applying geophysical tools can be estimated. It is important to know that when geophysics is used, it is likely to provide the correct description of the reservoir at the appropriate cost. Value-of-information (VOI) calculations (Figure 10) applied to seismic data acquisition, processing, and interpretation are often valuable to assure the oil and gas operator (and the geophysicist) that the right tools are being used in the right way.

**Figure 7.** Several types of tools are used to communicate uncertainty, including deterministic models that show the propagation of waves in the subsurface. Here, the model shows that surface seismic data will not illuminate some parts of the horizon under salt, resulting in a "shadow zone." In this model, many reflections bounce up to the base salt and become trapped because of conversion to refractions (Abriel et al., 2004).

**Figure 8.** Original oil in place (OOIP) is calculated from five critical parameters. The uncertainty in the parameters makes for a wide range of hydrocarbon estimates.

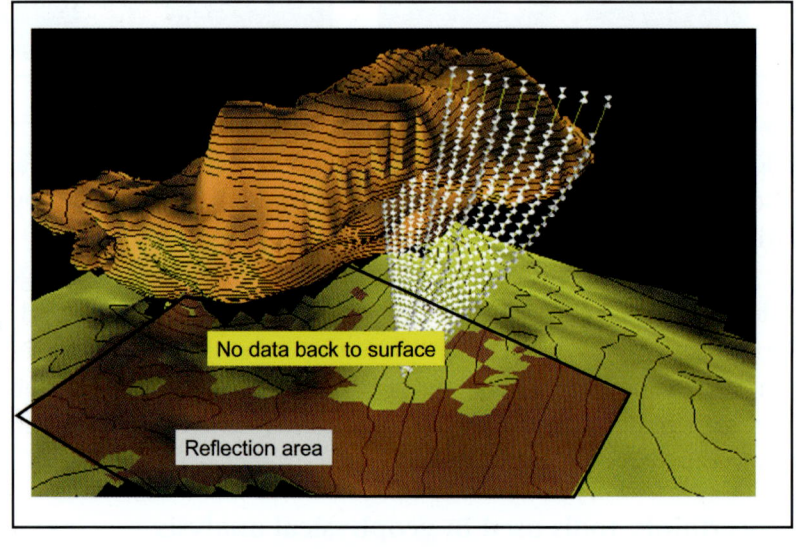

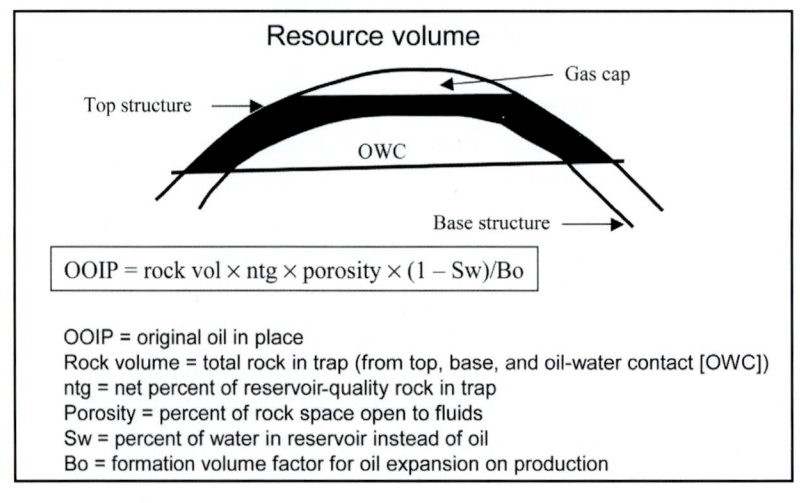

$$\text{OOIP} = \text{rock vol} \times \text{ntg} \times \text{porosity} \times (1 - S_w)/B_o$$

OOIP = original oil in place
Rock volume = total rock in trap (from top, base, and oil-water contact [OWC])
ntg = net percent of reservoir-quality rock in trap
Porosity = percent of rock space open to fluids
$S_w$ = percent of water in reservoir instead of oil
$B_o$ = formation volume factor for oil expansion on production

## Measurement of project value

To better appreciate the point of view of the operator, it is appropriate to review how projects are regarded from an economic perspective. With oil and gas operators, project value often is measured by total profit, cash flow, and net present value (NPV). NPV is a very strong driver in the hydrocarbon business, and we will see it applied throughout the life cycle of oil and gas projects. This economic measure attempts to take into account the future revenue derived from sales of production minus the cost of finding and the estimated costs of developing and producing the resource (Figure 11). NPV will be discussed in all sections of this text.

Although NPV is pervasive, another potentially significant economic driver for projects is "strategic fit." For example, a government might desire to secure a sufficient

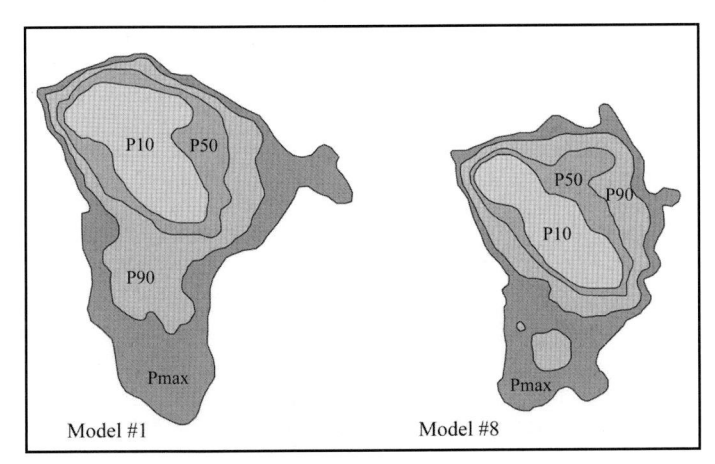

Figure 9. Parameters can be varied systematically or statistically to estimate the likely range of original oil in place (OOIP). Each OOIP realization is a specific combination of input parameters that constitutes a volumetric model. Each model has a range of likelihood for total fill, with the median known as the P50.

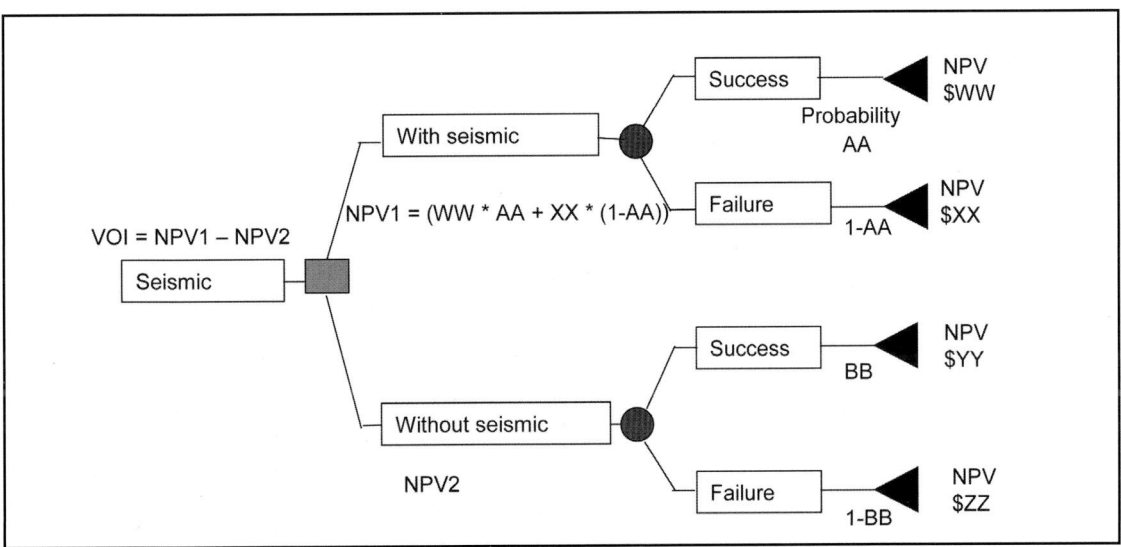

Figure 10. Value of information (VOI) is back-calculated from a decision tree that includes seismic data. The first branch on the left is a box in which one decides whether to use seismic data. If seismic data are used, the outcome branches from the circle, with results that might or might not be successful. The chances of success ("Probability AA") are estimated from experiences of experts. Estimated project values are the end result on the right. VOI then is back-calculated according to the formula shown.

energy resource for its economy, a source that will not be interrupted. That might result in a value for the project greater than the open-bid NPV evaluation, which assumes that other resources can be substituted through open-bid purchase. That is a point to remember when working with governments, national oil companies, and highly integrated international companies.

## Relative value of subsurface (seismic and nonseismic) information

What does one need to know about the subsurface to value the discovered resource? Primarily, the answers are: (1) the total resource size and (2) an estimate of the deliverability of production wells. It is necessary to identify the total area of the field, horizontally and vertically. Many fields are not purely structural traps, and stratigraphic boundaries often are encountered. In addition, many fields comprise multiple reservoirs that "stacked" vertically — reservoirs with different geologic characteristics and fluid content (Figure 12).

Net present value (NPV)

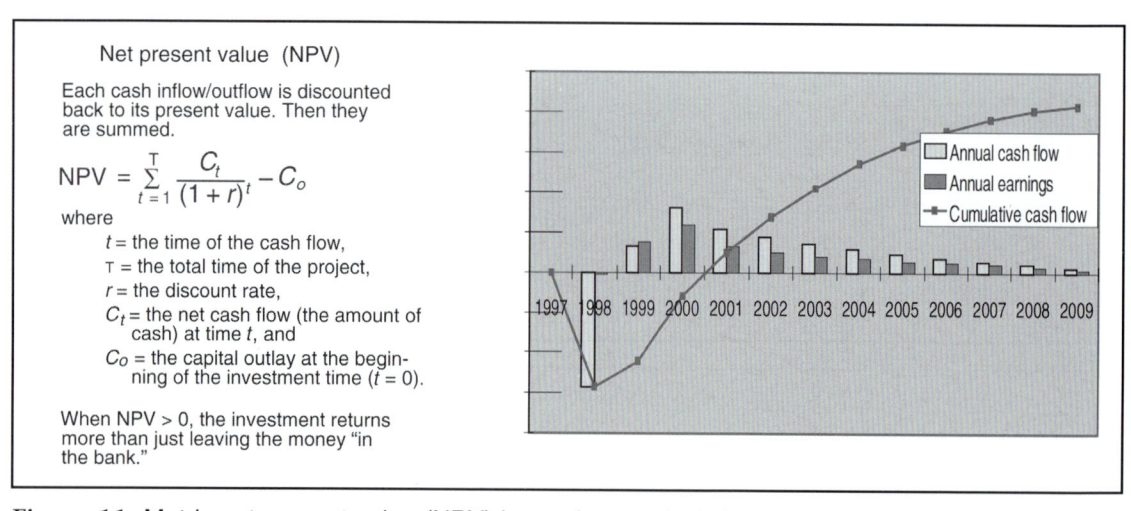

Each cash inflow/outflow is discounted back to its present value. Then they are summed.

$$NPV = \sum_{t=1}^{T} \frac{C_t}{(1+r)^t} - C_o$$

where
- $t$ = the time of the cash flow,
- $T$ = the total time of the project,
- $r$ = the discount rate,
- $C_t$ = the net cash flow (the amount of cash) at time $t$, and
- $C_o$ = the capital outlay at the beginning of the investment time ($t$ = 0).

When NPV > 0, the investment returns more than just leaving the money "in the bank."

Figure 11. Metric net present value (NPV) is used extensively in valuing oil and gas projects.

Figure 12. Oil and gas fields often have multiple accumulations that "stack" vertically. At least three reservoirs shown in this figure are represented by high amplitude. Vertical stacking of reservoirs is advantageous for development because fewer boreholes are needed to exploit the total resource (Sullivan et al., 2004).

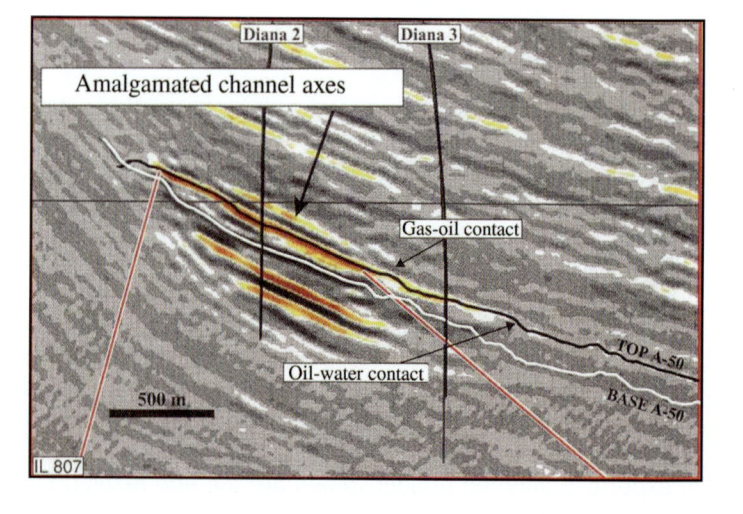

With exploration drilling success, the hydrocarbon recoverable estimate still has a wide range of subsurface uncertainty. An example of oil volume estimated from only one well plus seismic data shows how important it is to know the vertical and lateral extent of the reservoir (gross rock volume, or GRV) (Figure 13). In the example shown, GRV dominates the uncertainty. Because of that, a great demand is placed on analysis of seismic data away from the discovery well. To invest in a project, the uncertainty of hydrocarbons in place must be reduced, and reduction requires gathering information from expensive delineation drilling. Seismic data are a critical contributor to the process.

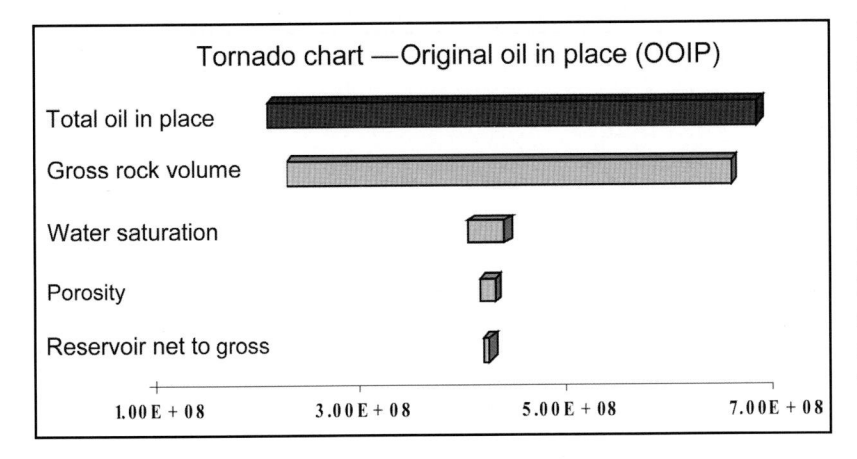

**Figure 13.** Key parameters contribute to resource estimates. The uncertainty of total oil in place (width of the upper bar) is the sum of the lower bars. The uncertainty of original oil in place (OOIP) in this case is caused mainly by uncertainty in gross rock volume.

# Chapter 2   Field Discovery and Delineation

## Reserves

Reserves in the subsurface can be classified as proven, probable, or possible (Figure 1). A discovery well is necessary to have proven reserves. Further drilling to reduce uncertainty in size and volume of the proven resource is an "appraisal" of the reservoir. In the same process or separately, wells drilled around or under the reservoir may prove up additional reservoirs that originally were considered to be probable or possible in a delineation process. Oil and gas operators do not completely agree on usage of the terms *appraisal* and *delineation*. To keep terminology consistent in this course, we will use the term *delineation* to cover both activities.

Prime measures of success during delineation are the determination of net present value (NPV) and the size of the booked reserves. The purpose of delineation is to establish the quantity and distribution of hydrocarbon reserves so as to plan the right scale of facilities for development and production of the field. The uncertainty range must be narrowed to reduce investment risk, but that process must be conditioned by the need to provide subsurface certainty as quickly and as cheaply as possible.

With enough delineation wells and flow testing, the estimate of the reserves can be established with limited uncertainty, but only at a cost. However, the cost of achieving low uncertainty is also detrimental to the project. The value of those same reserves is decreased with every delineation well and every additional expense of obtaining and using geologic and geophysical data. Time is also an element. If the project takes too long at start-up, the result will be a poor investment compared with the normal market return for investments.

## Geology of the exploratory well

What does the operator need to know about the subsurface during delineation, and how does geophysics help? Critical well data are gathered from the discovery well via downhole logs and rock samples from well cuttings. Reservoir parameters are tied to seismic traces and extrapolated to determine reservoir size and quality. From that, an estimate of the reserves is determined, and a program for delineation drilling is planned and executed. The delineation program must balance the need for information with the expense of obtaining it.

In the first step of planning a delineation program, seismic data are tied closely to reservoir properties derived from well data (Figure 2). In general, this includes all elements of the reservoir that describe the state of hydrocarbons in place — reservoir area, depth, thickness, stratigraphic model, net-to-gross thickness, porosity, fluid type, and saturation. The use of seismic data to estimate reservoir presence and properties away from wells requires that some geologic property known in the well (e.g., lithology) must express itself as a seismic attribute (such as amplitude). Then the seismic attribute is used as a proxy for estimating geologic property at localities away from the well (Figure 3).

**Figure 1.** Hydrocarbons in the subsurface are classified on the basis of quality of information on known quantity, flow capability, and marketability. The term *reserves* in the oil and gas business is accepted widely as covering the classifications of proven, probable, and possible reserves (adapted from SPE, 1997).

### Definition of reserves

Proved
- more than 90% chance that they are commercial at a given date forward under present conditions
- an area delineated by drilling and defined by fluid contacts
- generally requires reservoir production tests
- generally requires presence of facilities

Probable
- more than 50% chance that they will be produced
- anticipated to be proved by delineation
- increased production potential from enhanced recovery
- fault separated and lower than proven reserves

Possible
- more than 10% chance they will be produced
- could exist beyond probable reserves
- are in doubt commercially

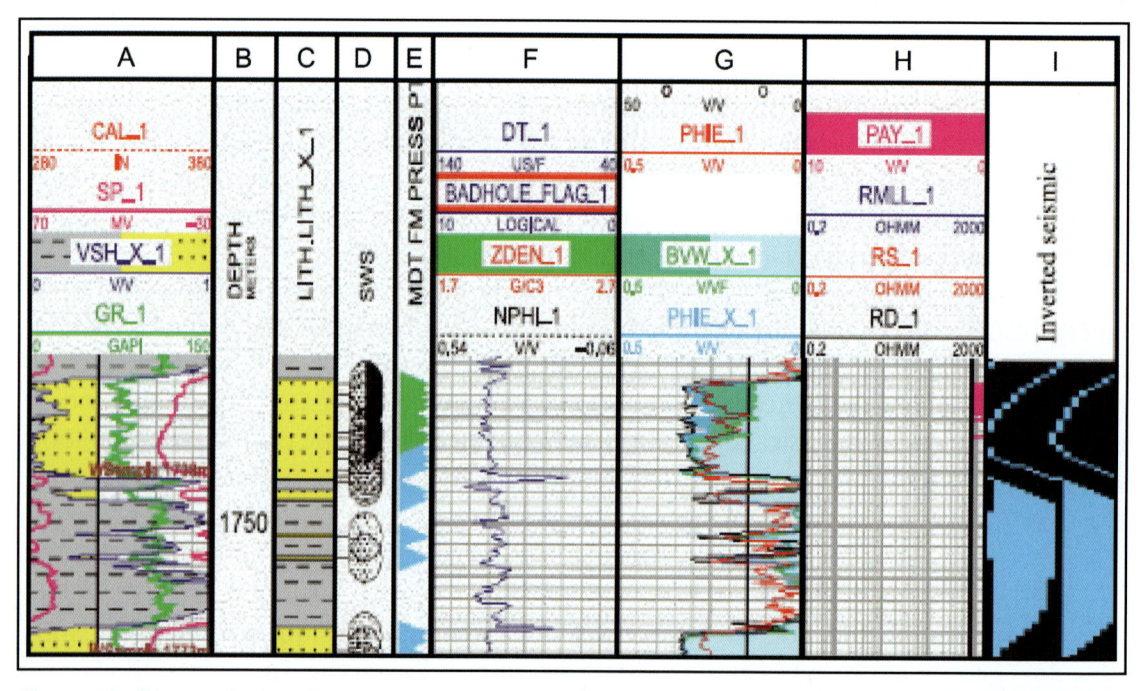

**Figure 2.** Reservoir data from the discovery well are derived mainly from well logs and bit cuttings. In column A, the significant information is the gamma-ray (GR) log curve, which is the basis for estimating the shale percent (Vsh, shown in gray). Column B shows depth. Column C represents the lithology calculation. Column D shows the location of sidewall samples (SWS); the half-black symbols represent oil shows. Column E shows information about formation fluid from the specialty sampling tool. Recovered oil is shown in green. Column F is the velocity log measured in transit time (DT), which shows that the sandstone reservoir has properties much like those of the shales. The density log (not pictured) has the character of the lithology log. Column G gives the response of the porosity log, showing high porosity in the reservoir sand. Column H shows assigned pay. Column I shows the seismic response, with an attribute to be described later, called inversion. Note the strong correlation of seismic data to lithology. Also note the difference in resolution of seismic and log data (Liu et al., 2004).

The impact of geology on the ability to recover reserves is very large. When production wells are to be completed, it is also critical to know what fluid will be produced, what area will be drained, rate of production, and duration. Those are the reservoir characteristics of well deliverability, which depend highly on geology.

An example of the impact of geology on total recovery in sandstone reservoirs implies that different depositional environments might require much different development strategies (Figure 4). Sandstone depositional environments that have been worked extensively by waves and tides generally have much better total interconnected porosity with fewer rapid changes in facies, which often are barriers to flow.

Dynamic information concerning the flow of hydrocarbons such as flow rate or flow properties such as permeability generally is not linked directly to seismic response, so it normally is not extrapolated with seismic traces. However, there are some specific and exciting exceptions to this rule.

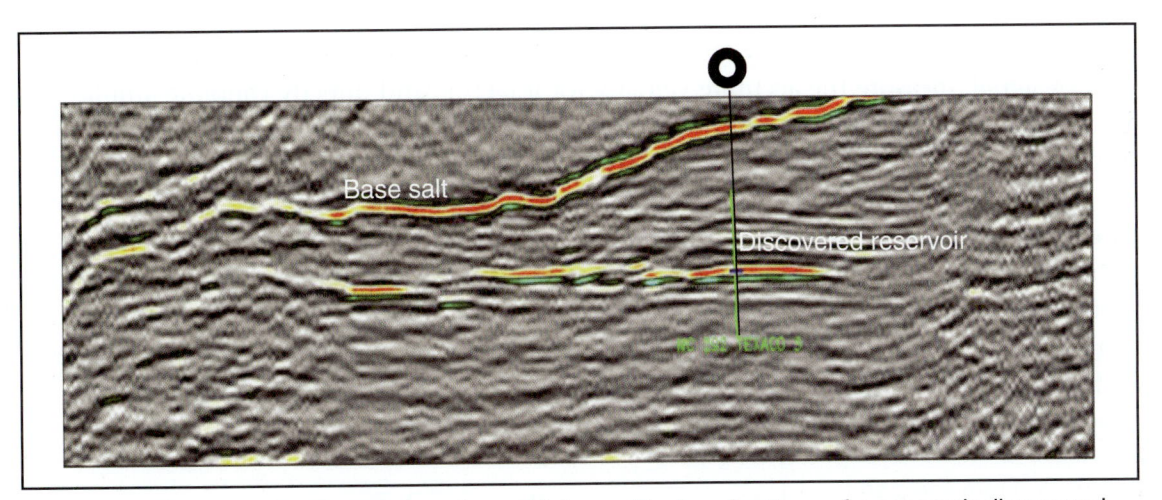

Figure 3. This seismic cross section shows high-amplitude reflections of a reservoir discovered below a salt body. High amplitudes (red and green) extend away from the borehole and can indicate the presence of hydrocarbons.

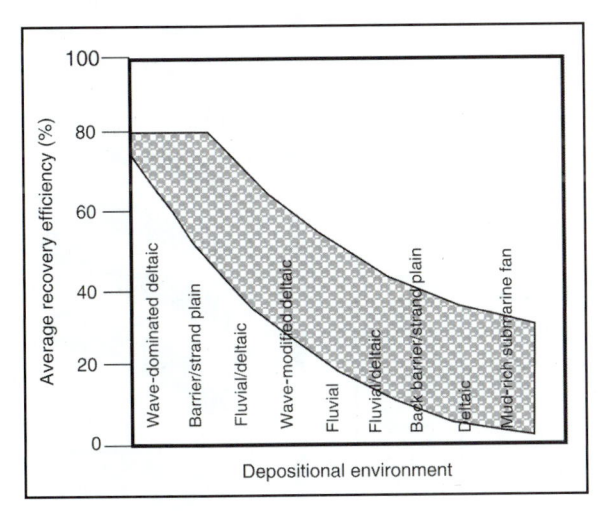

Figure 4. Hydrocarbon recovery is a function of geology. Reworked sandstone reservoirs have the best recovery efficiency, but total recovery has a wide range for each depositional environment (Larue and Yue, 2003).

## Correlating the well data to the seismic

Well properties that are to be extrapolated are tied to seismic data by using synthetic seismic models derived from sonic and density well logs (Figure 5). This allows for an understanding of how to relate seismic data, which are in time, to well data, which are in depth. It is important to know exactly where in the seismic data to locate the reservoir. Making the wrong tie and tracking the wrong seismic response would give false information, which can cause critical failure in reservoir evaluation.

Acquisition of vertical seismic profiling (VSP) at the well can provide the exact well tie required. VSP is a series of recordings from sources at the surface into receivers placed in the borehole. The exact tie of time and depth then is established.

If the sources are located at different surface locations, the VSP recordings can measure earth reflections close to the borehole at higher resolution than surface-to-surface recording. That can provide ground truth for earth acoustic parameters. When parameters are known exactly, surface seismic data can be processed better (Figure 6). The example shown uses VSP data to identify and eliminate multiples, measure and handle attenuation of signal, measure the accurate velocity for imaging, and directly calibrate amplitude variation with offset (AVO) for lithology and fluid discrimination. Ground-truth VSP data are used as the imaging target for reprocessing surface seismic data.

An additional point needs to be made about seismic-to-well ties. The well is only one data point in the reservoir. When using data from the well, it is unreasonable to believe there are no horizontal geologic variations in the reservoir. Changes in presence, thickness, and properties of local geology cause changes in the seismic observations, which must be modeled to be understood. By modeling an expected range of responses of the reservoir and surrounding rocks, it is possible to better classify measurements made from seismic data at localities distant from wells (Figure 7).

**Figure 5.** From sonic and density logs at the discovery well ($V_P$ and $\rho$), a synthetic seismic trace is generated and matched to the surface seismic data (far right). The synthetic trace is shown in red on the overlay. Seismic traces of the recorded surface data are in black, except for the trace at the well location, which is shown in green (Gratwick and Finn, 2005).

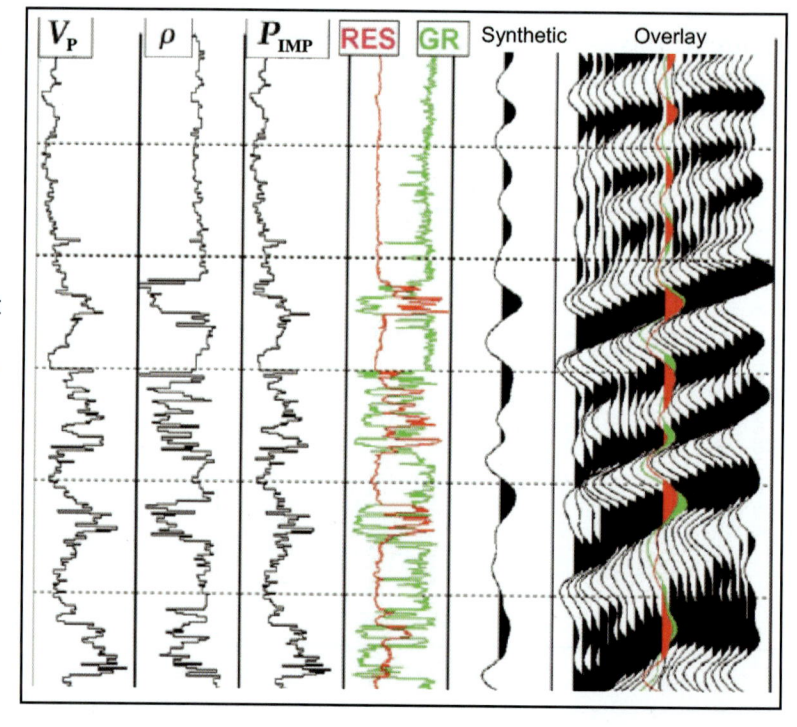

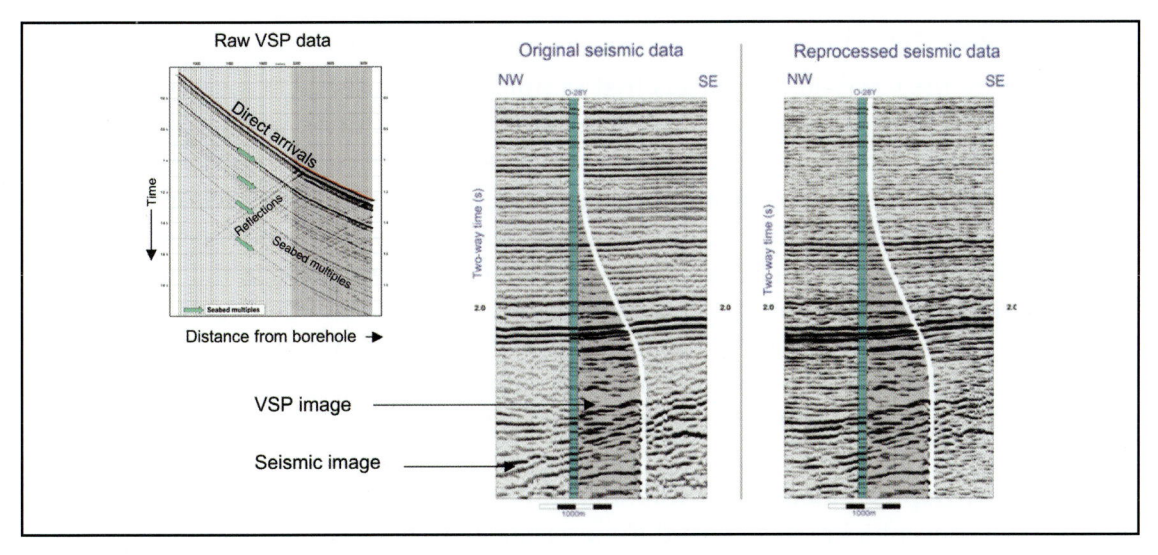

**Figure 6.** A raw recorded data shot of vertical seismic profiling (VSP) from the surface into the borehole shows direct arrivals, near-borehole reflections, and seabed multiples. When processed into a geometric section such as surface seismic data, it can be spliced into that section for comparison. With VSP as the ground truth, the processing parameters for surface seismic data can be adjusted to optimize imaging (Kaderali et al., 2007).

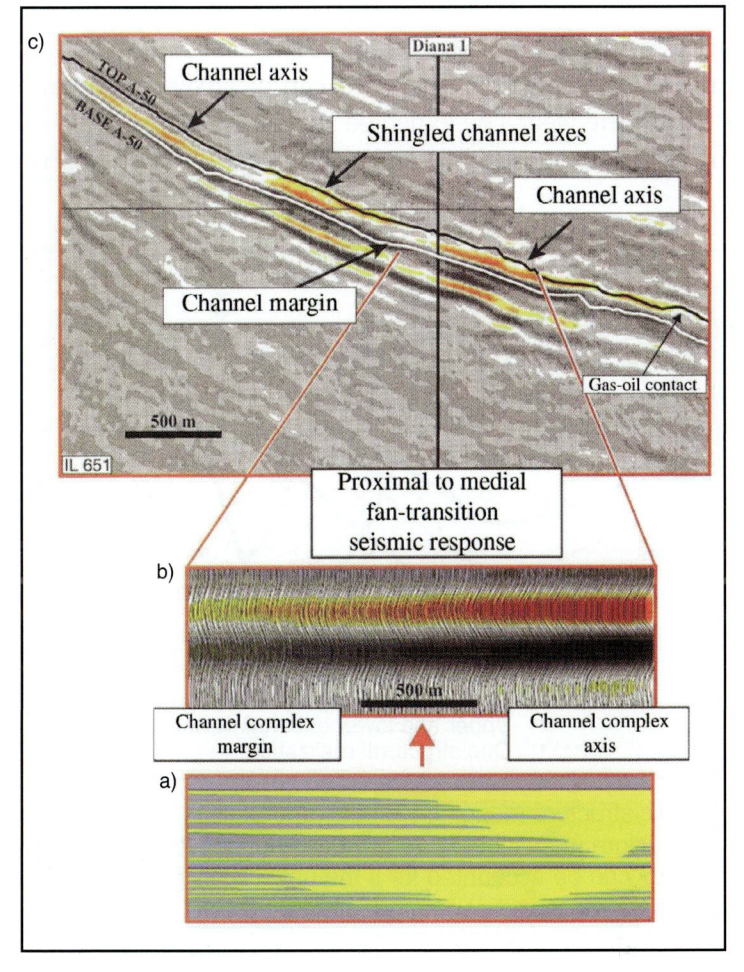

**Figure 7.** Forward models of seismic data are needed to understand the seismic response of the reservoir and fluid properties at the discovery well. The expectations of lateral variations in stratigraphy depicted in (a) are modeled to show (b) the potential seismic expressions. The pattern of variations can be matched to live seismic observations (c). This allows for an interpretation of where and how the reservoir varies (Sullivan et al., 2004).

## Extrapolating well data

What are the reservoir parameters that seismic data best reveal at localities away from wells? The most common are presence and structural position of reservoirs. Presence is established when the expected seismic signature of the reservoir is traced back to the well. Structural position is determined by identifying the correct reflections for the reservoir top and bottom at the well location and then mapping them. This is done in time domain and then the maps are depth-converted. However, structural depth away from the well is an uncertainty because of imperfect velocities also derived from seismic data. The total volume estimate of the reservoir is affected by uncertainty about depth; this is especially important in low dip areas. To help measure uncertainty about volume, a range of depth models of the reservoir away from well control should be considered (Figure 8).

The stratigraphic model is also important in the seismic extrapolation of well information. The conceptual model derived from seismic data prior to discovery is modified by data collected at the well. The reservoir then can be described better in terms of depositional environment and lithology. The conceptual model is used for understanding the expected size of the reservoir, the lateral change in properties, and the potential for compartmentalization (Figure 9).

Finally, estimates of fluid content also need to be extrapolated. If fluid contacts exist in the discovery well, they normally are projected over the closed reservoir volume. Often, however, some questions are unanswered by a discovery well, such as "Is there a gas cap?" or "How large is the oil rim?" At times, such questions can be answered in part by fluid modeling of seismic data tied to the well. The expected response of the fluid change then can be matched to the seismic response for predicting lateral fluid changes (Figure 10).

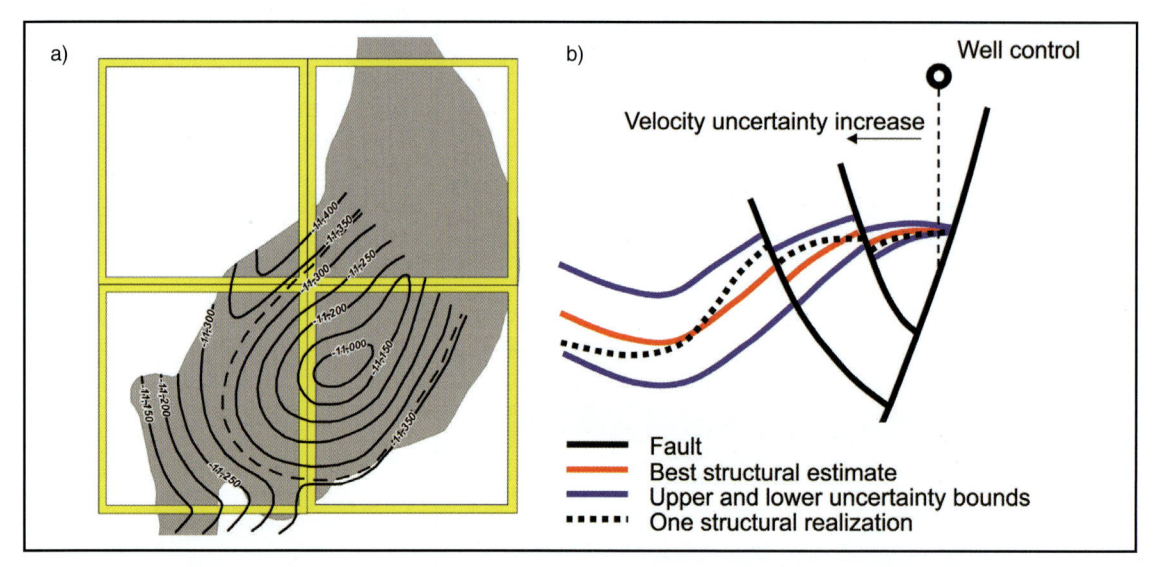

Figure 8. Time-depth conversion of seismic-structure mapping includes uncertainty of the structure away from the discovery well. One structure map, labeled as (a), does not convey uncertainty. (b) Depth uncertainty away from well control can be accounted for if the uncertainty of the velocity is measured. Different structural realizations then can be generated, all of which are equally likely. One depth realization is shown.

*Field Discovery and Delineation*

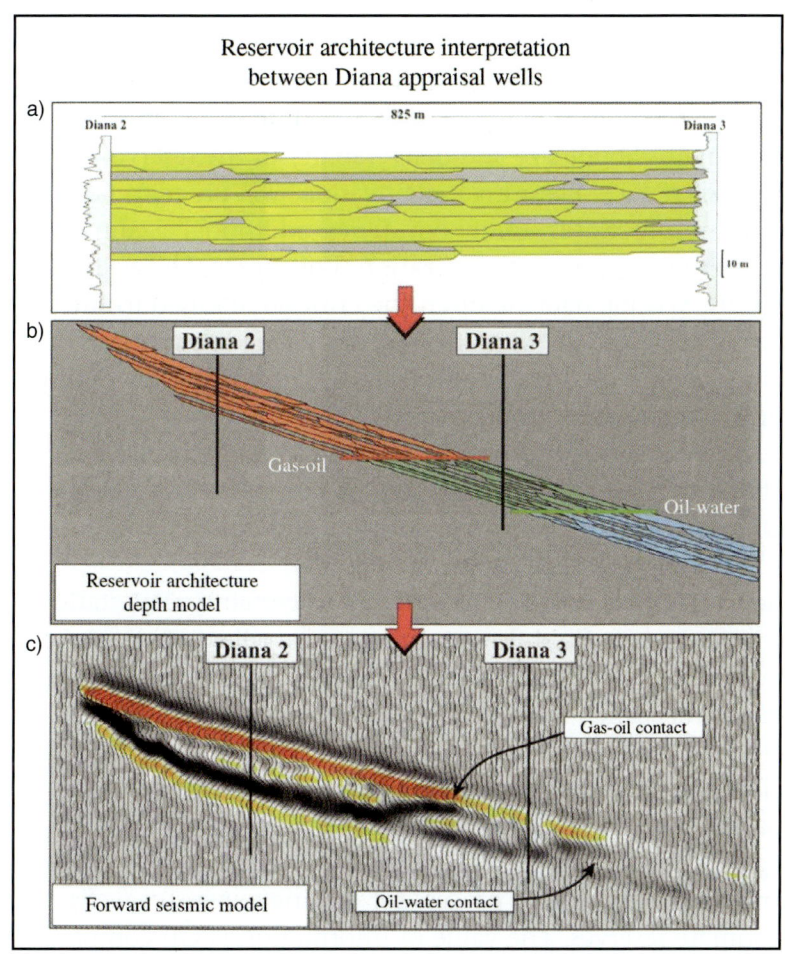

**Figure 9.** Extrapolation of well tie to seismic data is employed best by using a forward seismic model. Here, (a) sand units in two appraisal wells are modeled as a series of interconnected sand bodies. With (b) structure and fluid introduced, then (c) the effect on seismic response can be calculated (Sullivan et al., 2004).

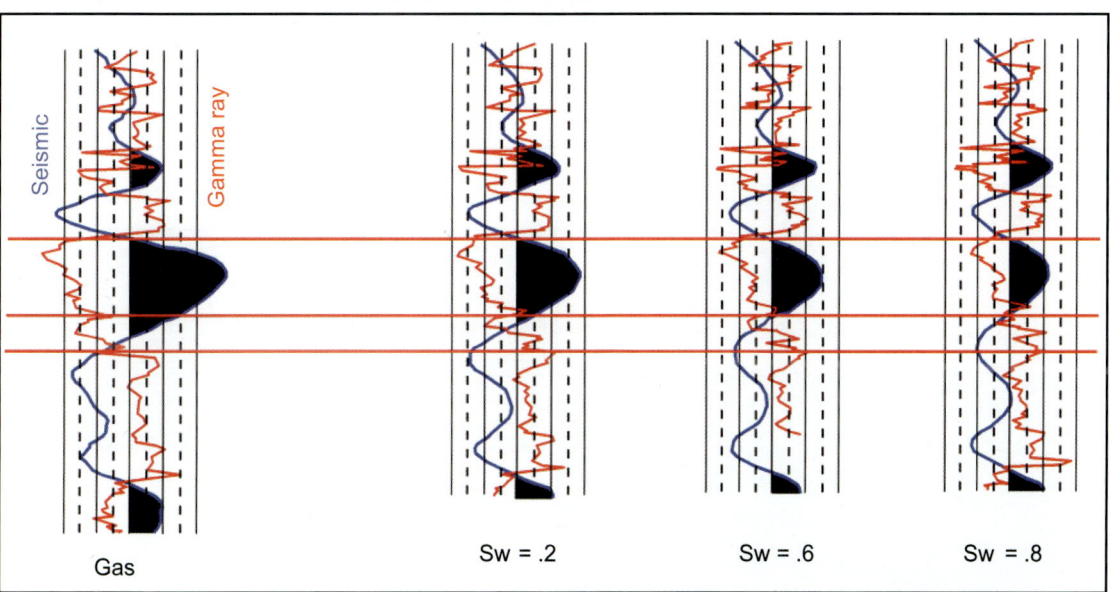

**Figure 10.** Seismic-response models of the expected variations in fluid also are generated from logs of the discovery well. Here, a large seismic response is obtained from a gas-filled reservoir. In contrast, oil with water saturation (Sw) introduced at 20%, 60%, and 80% shows a diminishing of the seismic response.

## Delineation strategy

After the exploratory well is drilled and tied to seismic data, the seismic attributes and geologic modeling serve as the foundation for new estimates of reserves. However, significant uncertainty usually remains about size and deliverability of the reservoir which only delineation drilling can reduce. Planning a delineation-well program takes into account the need to confirm resource area, fluid boundaries, additional accumulations, lateral variations in the reservoir, compartmentalization, pore space, and deliverability of the development wells. Seismic data are often the critical data used to site delineation wells. The accuracy of reservoir estimates from seismic data is important. If the seismic data are believed to correctly predict conditions away from wells, that eliminates the need for more expensive delineation drilling (Figure 11).

Some common tools are used during delineation to measure the importance and range of uncertainties. Uncertainty ranges of key parameters are compared (Figure 12). From that comparison, a plan is derived to reduce uncertainty in the most important parameters.

The static reservoir model is another tool used to estimate uncertainty. The static model, although not necessarily meant for fluid-flow simulation, can illustrate the best guess of the subsurface reservoir model as a digital object. When the static model is useful, the range of reservoir parameters (which is uncertain) can be sampled, and multiple reservoir models can be generated to understand the impact of uncertainty on the estimation of hydrocarbons in place (Figure 13). If necessary, midrange and end-member models can be selected and passed on to fluid-flow simulations as well.

In addition to measuring and illustrating the uncertainties about properties of the reservoir, it is important to formalize the decision process for drilling expensive wells versus using seismic data to estimate reserves. The value of information (VOI) comparing data from the well and the seismic data can be considered, and the trade-off of expense versus certainty can be measured. Seismic data are not always the obvious choice for information control (Table 1).

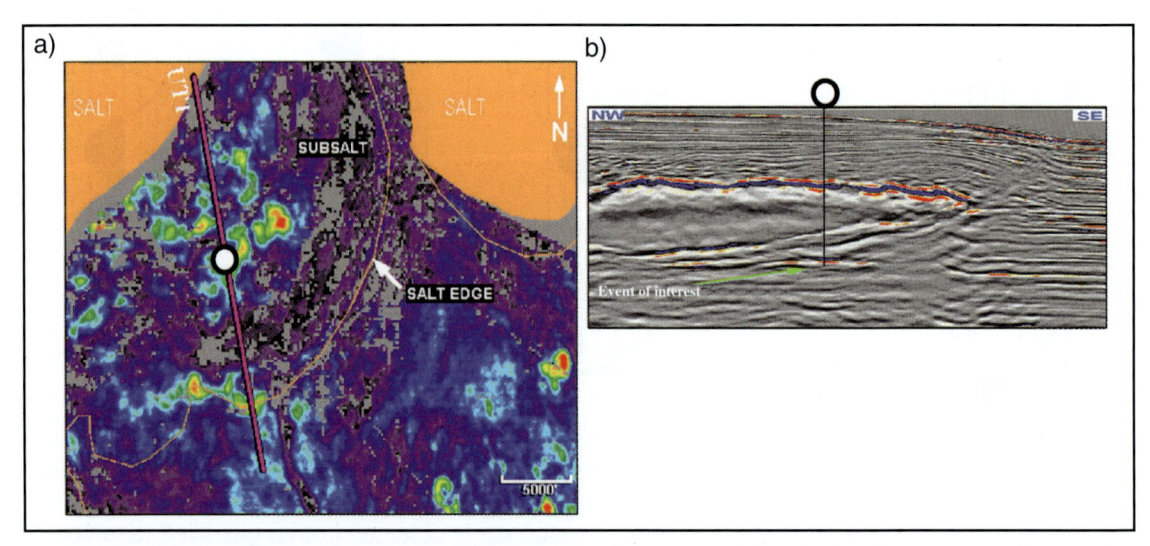

**Figure 11.** (a) The amplitude map and (b) the seismic section through the discovery well are primary tools for trying to find the limits of the reservoir (Abriel et al., 2004).

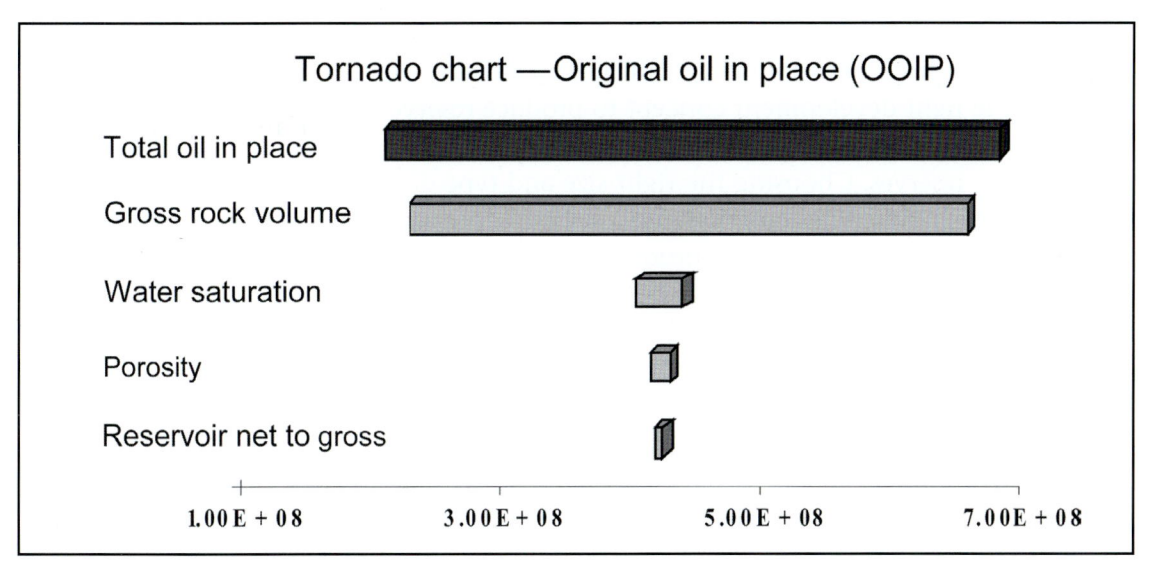

**Figure 12.** Uncertainty in the range of reserves can be influenced strongly by only a few parameters, as shown in the tornado chart. Note that uncertainty in gross rock volume is the largest contributor to uncertainty for oil in place in the example project. In the early stages of a project, uncertainties often are dominated by the unknowns in reservoir volume. Delineation drilling programs, therefore, are designed to confirm boundaries and major stratigraphic variations.

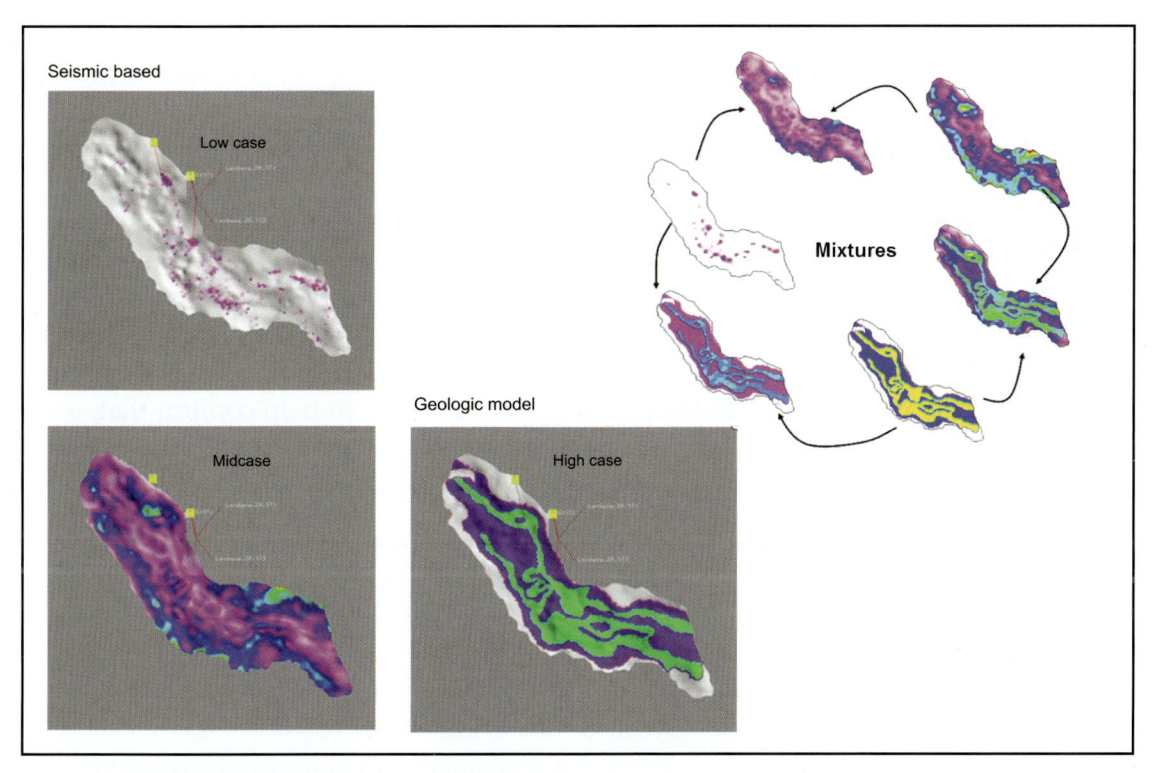

**Figure 13.** Reservoir models also can be used to measure and illustrate uncertainty in reserves before and after delineation drilling. Models can be built combining geologic concepts and direct seismic observations in different proportions. In the porosity case shown, the geologic model (high case) is more optimistic than the direct seismic observations (low case). Additional reservoir realizations using different proportions of seismic and geologic control are shown in the range marked *Mixtures*.

At the end of the delineation cycle, important decisions must be made about the viability of the project economics (go ahead and develop or sell now) and about the choice of the right development concept to produce reserves. Choice of development is highly dependent on surface access (desert, ice, distance from pipelines) and the nature of subsurface reserves. Choosing the right size and type of production facilities is critical to project value. If facilities are too large for the reserves produced, too much money will be spent. If facilities are too small, production is limited, and the value of the project is reduced because of delayed extraction. When choosing facilities for development and production, reservoir geophysics might contribute the greatest value.

## Case history

A useful case history helps to communicate the significance of seismic data to delineation and project value. For example, a discovery well in deep water was drilled on a high-amplitude anomaly that contained oil in a sandstone deposited in deep water (Figure 14). Seismic data were of such high quality that prestack traces used to image the reservoir were very stable and gave excellent measurements of variation with offset.

The basic principles of amplitude variation with offset (AVO) (Figure 15) require an understanding of the difference in reflectivity between normal incidence reflection (R0) and reflectivity at higher angles. It also is necessary to model the petrophysics of seismic response to different kinds of fluids in the reservoir. Then seismic data are measured for variation with offset, and parameters such as the gradient can be observed. Finally, the model and observations are compared to understand whether the AVO is measuring variations in fluid or lithology.

A successful discovery well was drilled into a sandstone that was oil full and located close to the water contact (Figure 16). The updip structure might contain more oil but also might have a significant gas cap, which would change the value of the project greatly. It was important to know whether the facilities to be placed in deep water should be designed for gas and oil or only for large volumes of oil. Three-dimensional seismic data were used to map reservoir structure, and a horizon slice was taken from a reflector at the top of the reservoir (Figure 17). The amplitude from the horizon slice is significant because the high-amplitude updip fits the model of a gas cap. To help confirm that, a 2D

Table 1. A comparison of factors shows why seismic data might be favored in some reservoir delineation areas more than in others.

| Deepwater clastics | Onshore carbonates |
|---|---|
| • if seismic responds well to reservoir variations | • if seismic responds poorly to reservoir variations |
| • if reservoir properties vary gradually in the geologic model | • if reservoir properties are unpredictable in the geologic model |
| • if lateral extent is in question | • if lateral extent is reasonably certain |
| • and wells cost ~$60 million | • and wells cost ~$10 million |
| • and seismic costs ~$20 million | • and seismic costs ~$20 million |
| • then saving one well easily pays for seismic investment | • then seismic is cost-challenged |

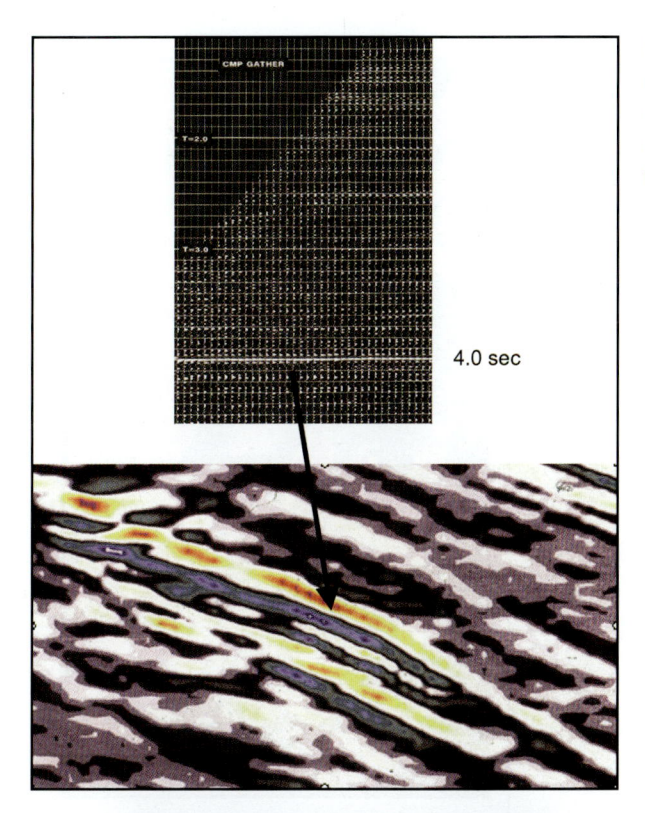

**Figure 14.** In this seismic expression of a discovery in deepwater turbidites, the high-amplitude reflection of the reservoir is seen also in the high-fidelity seismic response from the prestack traces at 4.0 s.

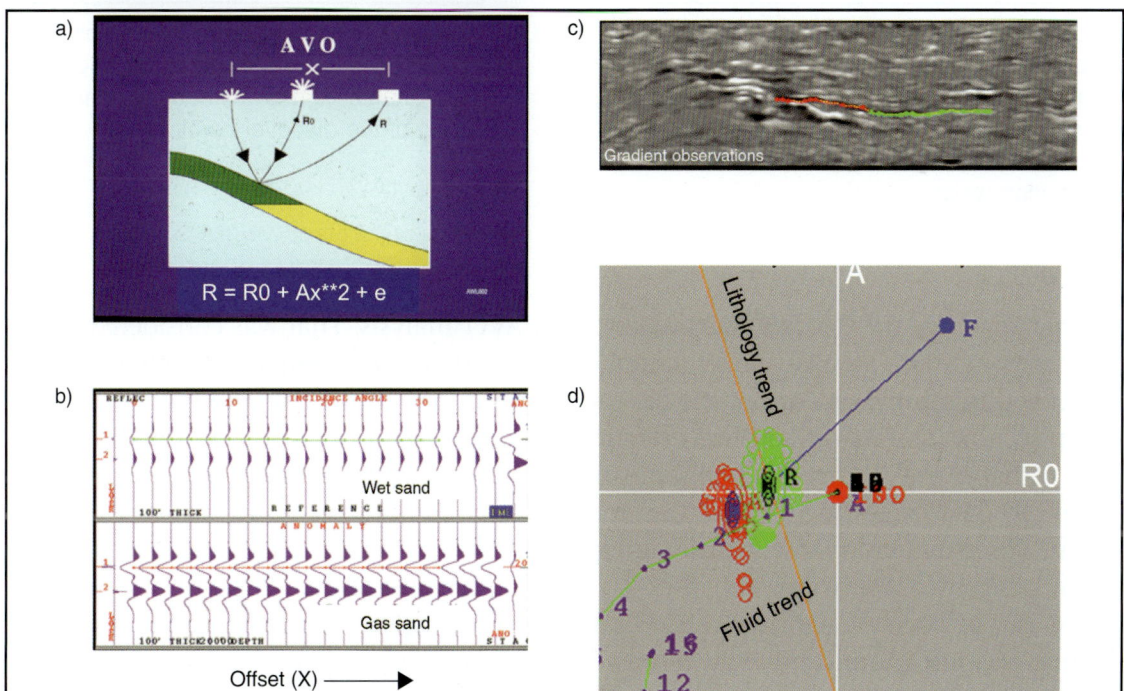

**Figure 15.** (a) The concept of AVO is that the reservoir is illuminated by waves that come at different angles and have different responses to the reservoir geology and fluid content. (b) Models of the seismic response of the reservoir with and without hydrocarbons show the magnitude of response to different fluids. (c) Seismic sections of gradient (shown) and zero offset are interpreted for the hydrocarbon (red) and water (green) zones. (d) Oil and water respond differently in AVO, as seen in this crossplot of the zero offset and the gradient.

*Reservoir Geophysics: Applications*

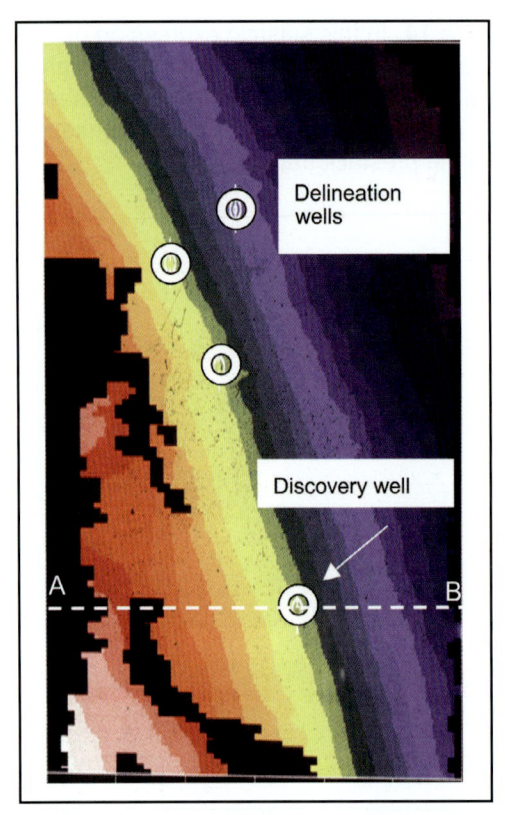

**Figure 16.** This structure map of the reservoir shows dip of less than 10°. The location of the 2D seismic line from the 3D survey is shown, along with the locations of the oil discovery well and three delineation wells.

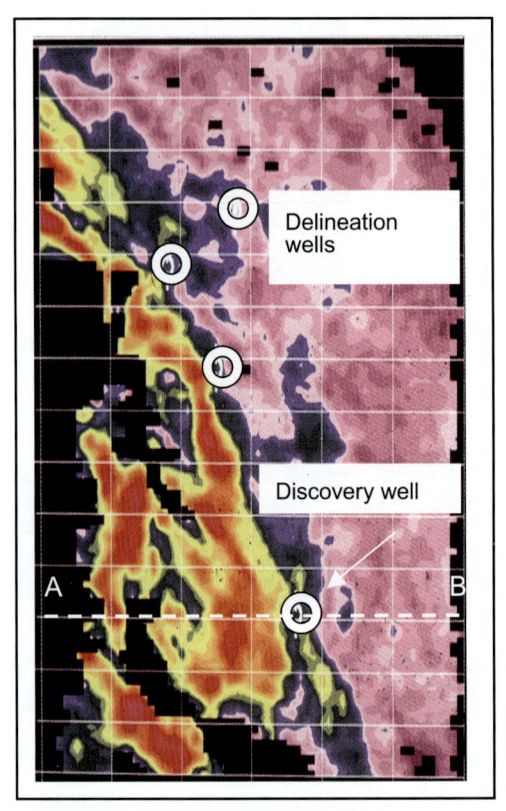

**Figure 17.** The horizon slice (amplitude map) of the top of the reservoir shows the variable high-amplitude response updip of the discovery well.

seismic line from the 3D data was processed for AVO analysis. That was considered to be a safe procedure because the structural dip is less than 10°. Comparing the near- and far-reflection seismic sections of the 2D results (Figure 18) shows that the high-amplitude updip section increases at far offsets (Figure 19), which again fits a model of updip gas.

If that is true, a large gas updip meant the project economics were in jeopardy. When 3D data processing for AVO was applied to the project, the result was very different. Three-dimensional seismic imaging attempts to collapse the recorded image to a point, whereas 2D imaging collapses the data to an ellipse (Figure 20). The size of the ellipse can be large enough that the 2D imaging is "smearing" or averaging the reservoir seismic response significantly and can give an inaccurate or false answer even when there is no structural dip.

In the case shown, the 2D and 3D seismic responses of migrated data and the equivalent gradients are very different, and the 2D seismic data give a wrong answer (Figure 21). A horizon slice of the AVO attribute that depicts decreasing AVO in blue and increasing AVO in red clearly shows the 2D AVO prediction to be in error (Figure 22).

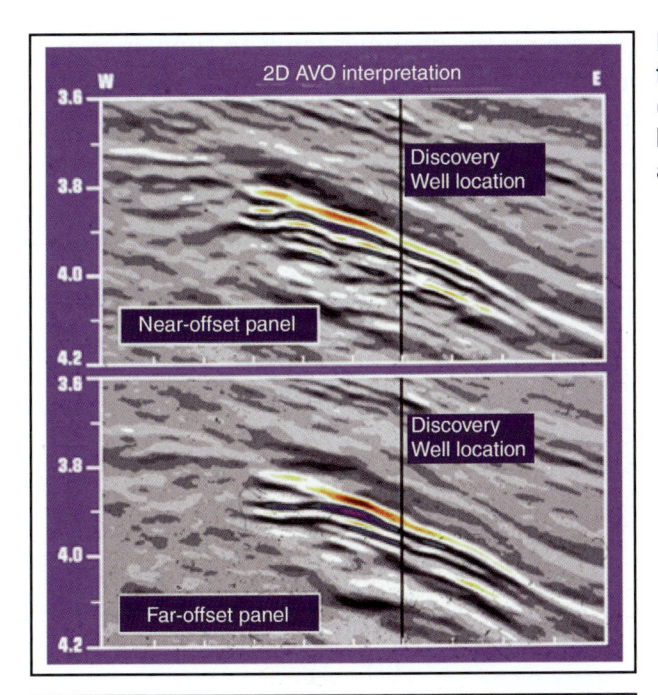

Figure 18. The 2D seismic line processed for AVO shows the near-offset amplitudes (top panel) updip of the discovery well to be lower in amplitude than the far-offset amplitudes (bottom panel).

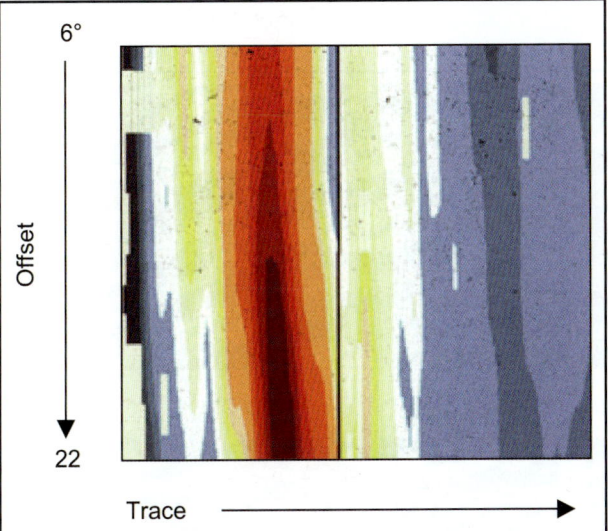

Figure 19. The value of the amplitude along the horizon for near offset is at the top of the plot and for far offset is at the bottom. The increase in amplitude with offset is shown by the increase in value vertically down the plot.

Instead of the increasing AVO suggesting a gas cap, the area just updip of the oil discovery is a deep negative AVO. No evidence of a gas cap could be supported; the focus of the project turned back to a situation of oil filling the reservoir. Subsequent delineation and production wells have confirmed that.

Having a false indicator from 2D AVO was not helpful, and the lesson of using 3D imaging for AVO was pursued from then on. But the interesting question remains, "What causes the high variation in AVO seen on the horizon slice?" The answer is found in the model of the stratigraphy. The amalgamated channels of the deepwater gravity sand deposits can be separated into two facies — the channel and the overbank. Channels tend to be thick, blocky, and sandy, whereas overbank deposits are laminated sandstones with interbedded silts. The AVO response for the channels tends to show increase in amplitude with offset, whereas the overbank deposits decrease (Figure 23). Thus, the AVO horizon slice in this case is related primarily to the facies type and is independent of fluids.

**Figure 20.** The subsurface response to seismic data is not useful unless the data are migrated. Three-dimensional migration focuses the data from a wide circle into a smaller one limited by the frequency of the propagating wavefield. Two-dimensional migration only focuses the data inline into an ellipse, which leaves a significant crossline amplitude averaging of the subsurface.

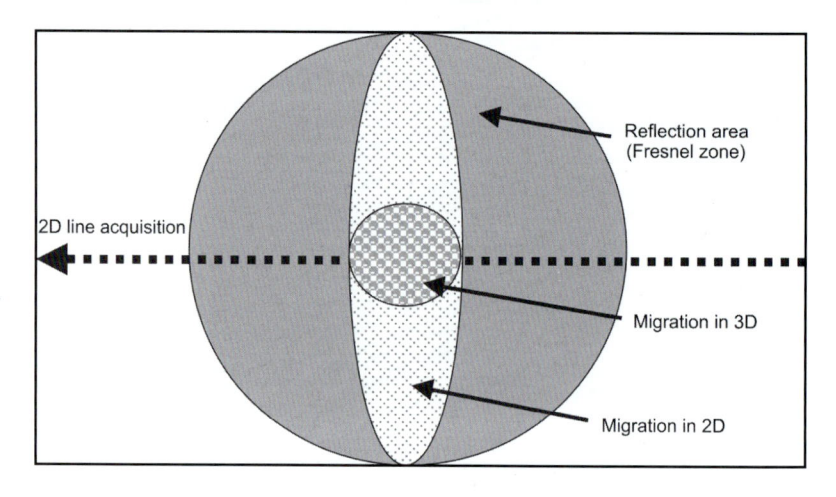

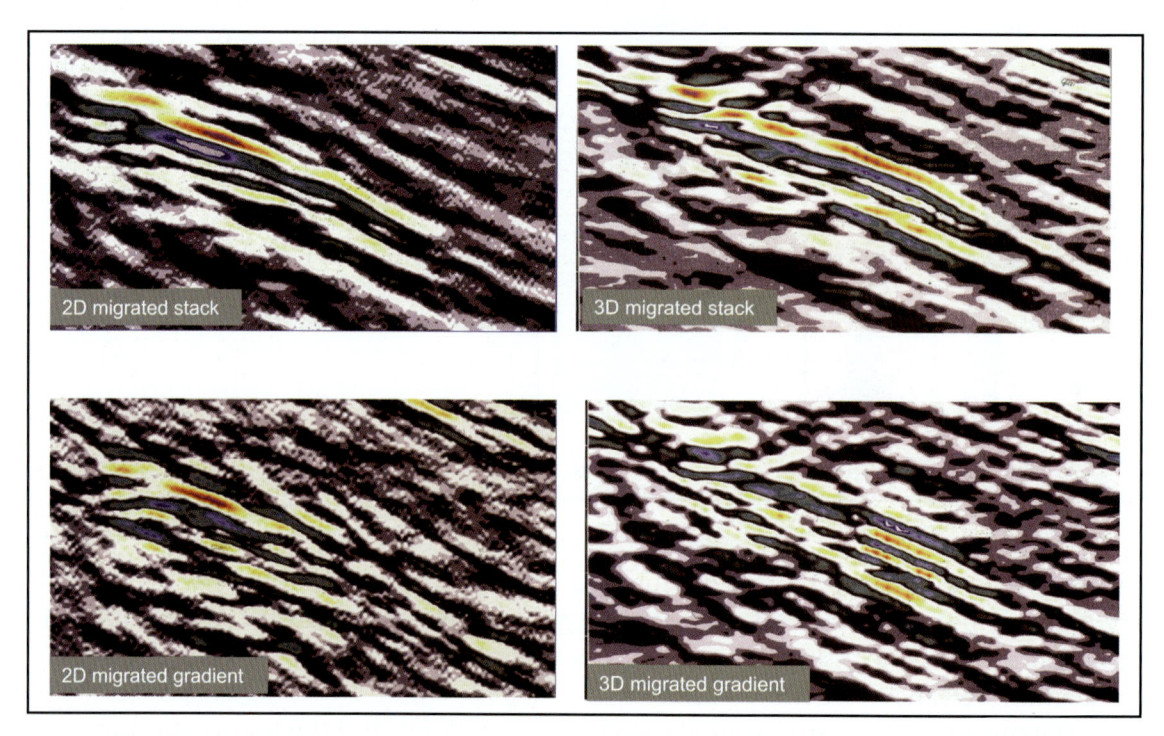

**Figure 21.** The processing of these data was exactly the same except for 2D and 3D effects. Reservoir response of the 3D stacked and migrated data compared with 2D is different in the details of reservoir amplitude. The gradient calculation is even more sensitive to 3D effects, as this comparison shows — 2D gradient calculations give the wrong answer.

This is a lesson we need to keep in mind. Although we may desire attributes such as AVO to discriminate fluids in the reservoir directly, AVO primarily responds to changes in rock type. We cannot make predictions about fluids without first having a valid and accurate geologic model.

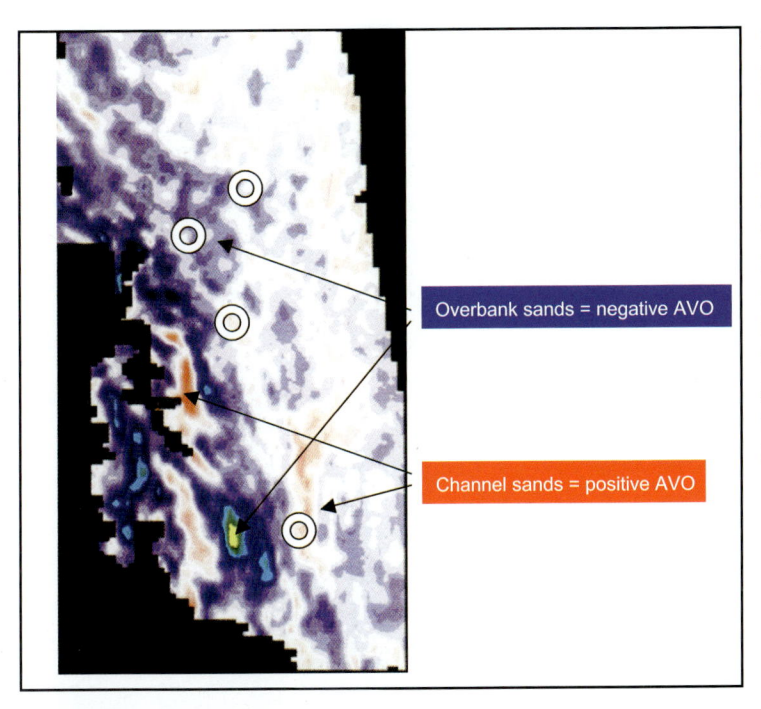

Figure 22. The AVO attribute of restricted gradient is the absolute value of the increase in offset. The horizon slice of this attribute shows that the reservoir directly updip of the discovery well is not positive AVO (red), as 2D data indicated, but rather is a strong negative (blue). The AVO is responding primarily to stratigraphy. Positive AVO anomalies are associated with channel facies and negative with laminated sand facies.

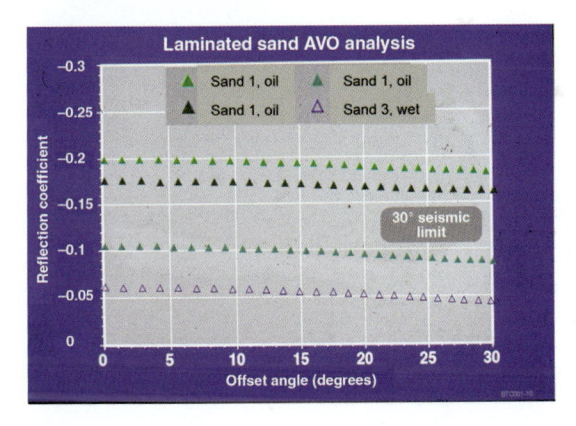

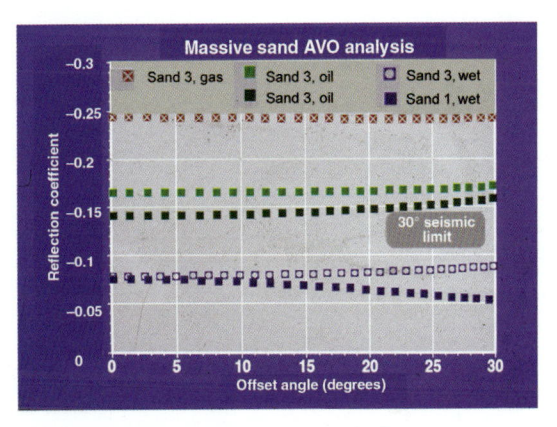

Figure 23. In this comparison of AVO measured from well data for two geologic facies, well data confirm that absolute amplitude is influenced strongly by fluid content. Oil and gas give much higher reflection than water-saturated reservoir sands do. Well data also confirm that for this example, AVO is primarily a function of stratigraphy. The AVO response decreases in laminated sands and increases in channel sands.

# Chapter 3  Delineation Problem

### Prologue

The problem presented below was a project initiated 10 years in the past, so we can present the problem and see the resolution. In the oil and gas business, it sometimes takes an entire decade to find out if your ideas about a reservoir are correct. For the present-day operator, the technology mentioned below will seem up to date, but the operating costs and product prices might be low according to today's standards.

### Objective

Working with a delineation team is an exciting and rewarding experience. Just after discovery, subsurface work is focused on sufficiently determining the volume and extent of the hydrocarbons proven in the discovery well. Geophysics has a strong role to play in this effort, and the following problem is designed to include you in this type of experience. Here is what you must do:

1) Given key information concerning a case-history discovery that requires delineation, define where to drill delineation wells.
2) Given a list of applications of geophysics, define a geophysical program to improve the subsurface interpretation.
3) Discuss the delineation program design and the reasoning.
4) Realize that there is no "right" answer to the problem, but the actual results of the delineation problem will be shown.

### Key background information

You have just joined the deepwater team and have been put on the Salinas project, which had been started in 1997. Lucky you! The exploration department discovered gas in reservoir sandstone at the 103 #1 well. The section of the well log with the discovered gas (Figure 1) shows the sandstone to be of low resistivity and highly laminated. The structure map (Figure 2) illustrates that the well was drilled on a structural high under a salt-dome overhang. No water contact was seen in the well. It is important to find out how large the gas accumulation is and whether it will be economic. It might be filled to spill point. It also might be only partly full or have serious stratigraphic restrictions.

The key seismic line through the well (Figure 3) shows why it was drilled. An amplitude anomaly is below the salt, but it is discontinuous. Because the subsalt imaging is imperfect, it is not known whether the reservoir is continuous or even whether it extends laterally from underneath the salt. The horizon slice (amplitude map) of the reservoir interval (Figure 4) shows clear stratigraphic information only when the horizon emerges from under the salt. Beyond the salt margin, the reservoir appears to be related to a sand-channel system.

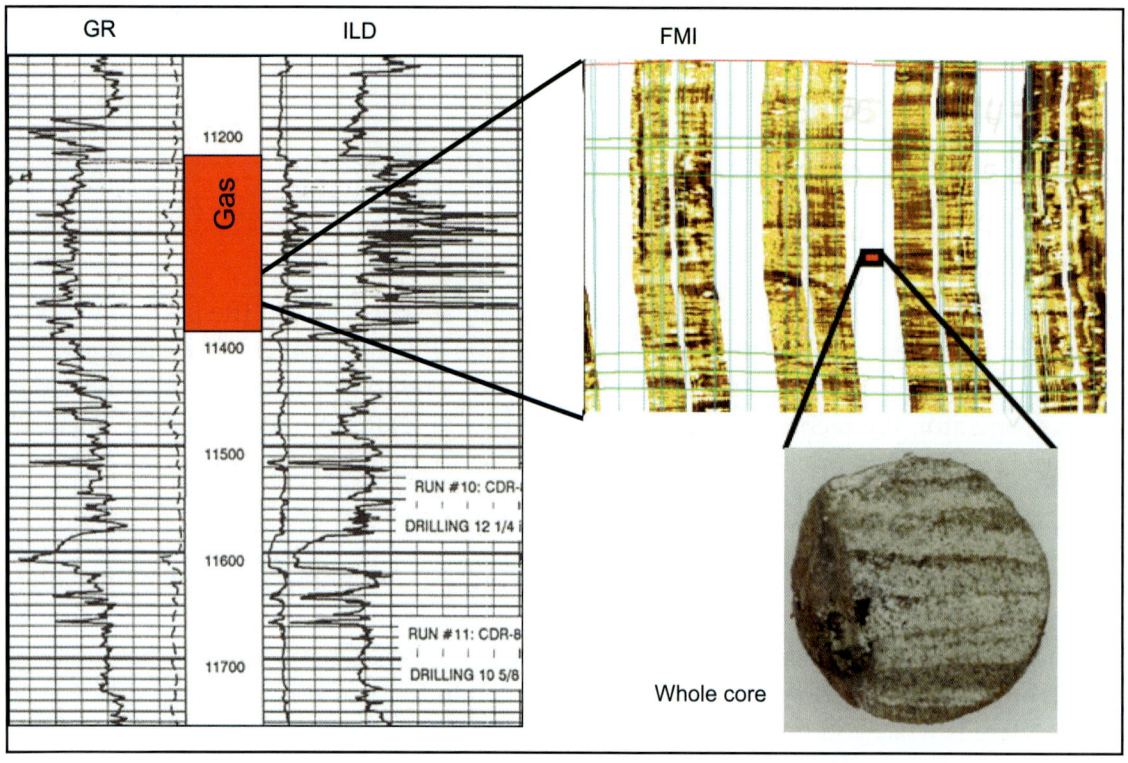

**Figure 1.** The discovery well contains gas in a laminated deepwater sandstone. Data collected from the well include wireline logs as shown here, sidewall cores, conventional core (shown), formation microimager (FMI), which is a microresistivity log (shows where core was taken), and flow tests. Note that the depth on the log is referenced to the Kelly bushing 85 ft above sea level.

**Figure 2.** The structure map of the reservoir in depth and referenced to sea level shows the location of the discovery well. A large salt body exists to the north and west, so reservoir is not present there. The well is gas full with no water contact. The projected water contact estimated from reservoir pressure is shown as a dashed line.

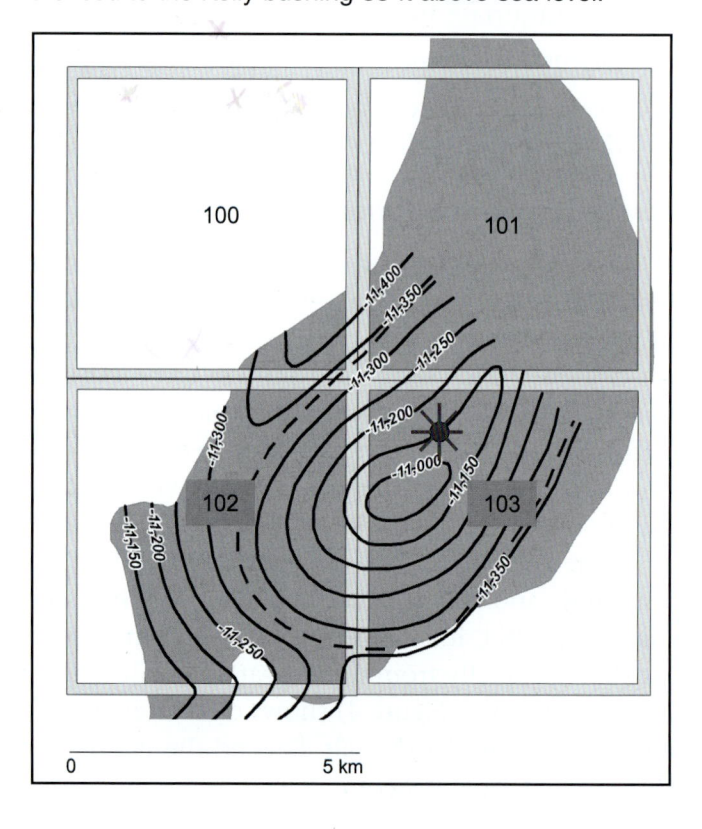

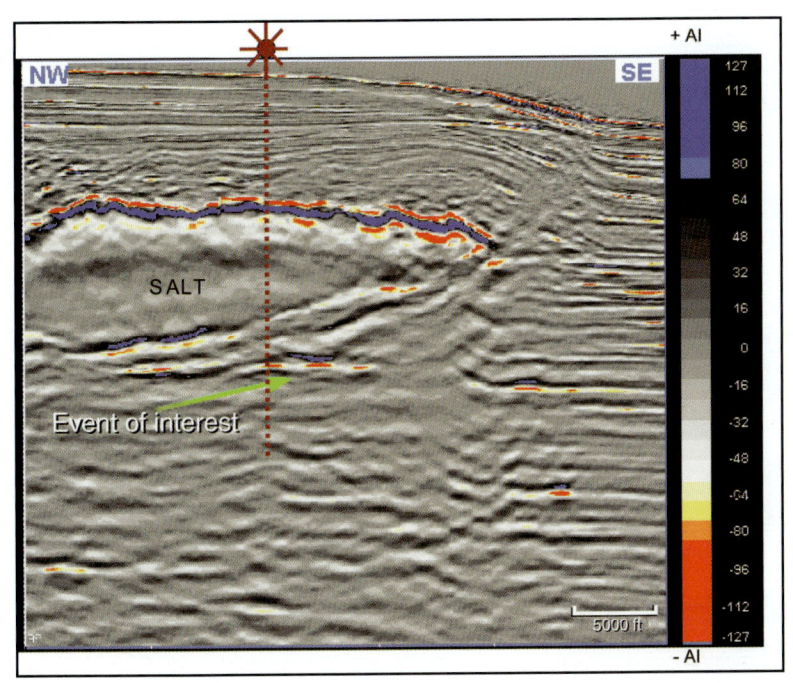

**Figure 3.** A poststack depth-migrated seismic cross section shows that the discovery well was drilled into an amplitude anomaly below salt. The seismic response of the reservoir is discontinuous laterally. Polarity of the data is low impedance, shown in red. Some caution in the use of data is advised because, in comparison to data from areas beyond the salt, the subsalt seismic data appear to be poorly focused.

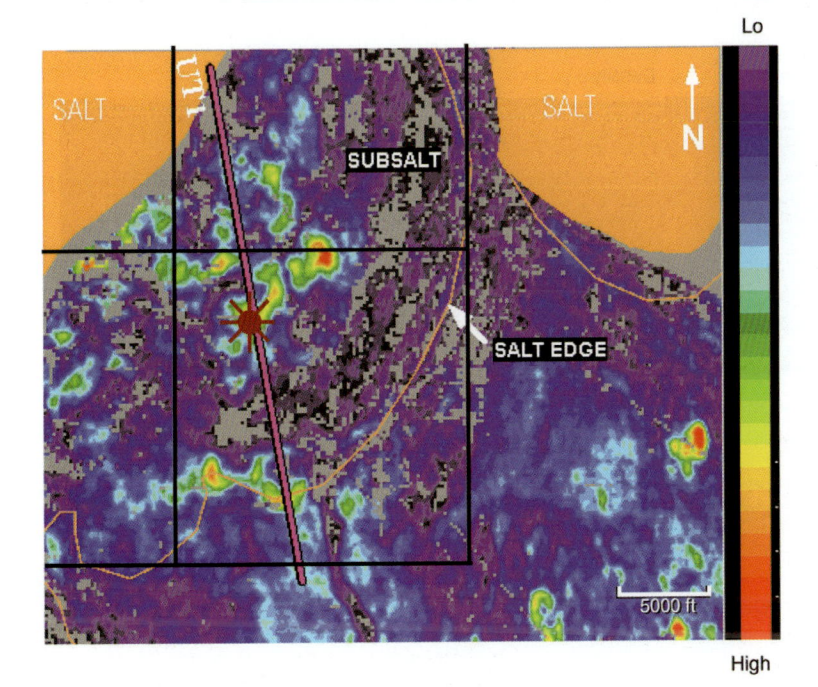

**Figure 4.** The horizon slice at the reservoir level shows that high amplitudes (yellow and red) are discontinuous in areas beneath the salt. Outside the area of the salt, amplitudes show evidence of a sand-channel system.

The conceptual geologic model (Figure 5) suggests that the main channel system was initiated and cut during falling sea level and filled with semicontinuous and interconnected channel sands with silty overbank deposits. During a rise in sea level, the deposit was cut by a final channel that might be filled with shale. As measured from logs of the discovery well, the geophysical response of the sand varies significantly with porosity (Figure 6). Therefore, the amplitude map (when perfect) is a good proxy for the presence and quality of the reservoir. It is possible that the sand does not extend over the entire structure but is restricted to the east and west by pinch-outs.

**Figure 5.** The stratigraphic model of the sand-channel system shows that the sand was deposited in a deepwater amalgamated channel system primarily during a sea-level lowstand. The individual sand bodies are limited in extent, but many of them cut into one another, composing a partly connected flow unit.

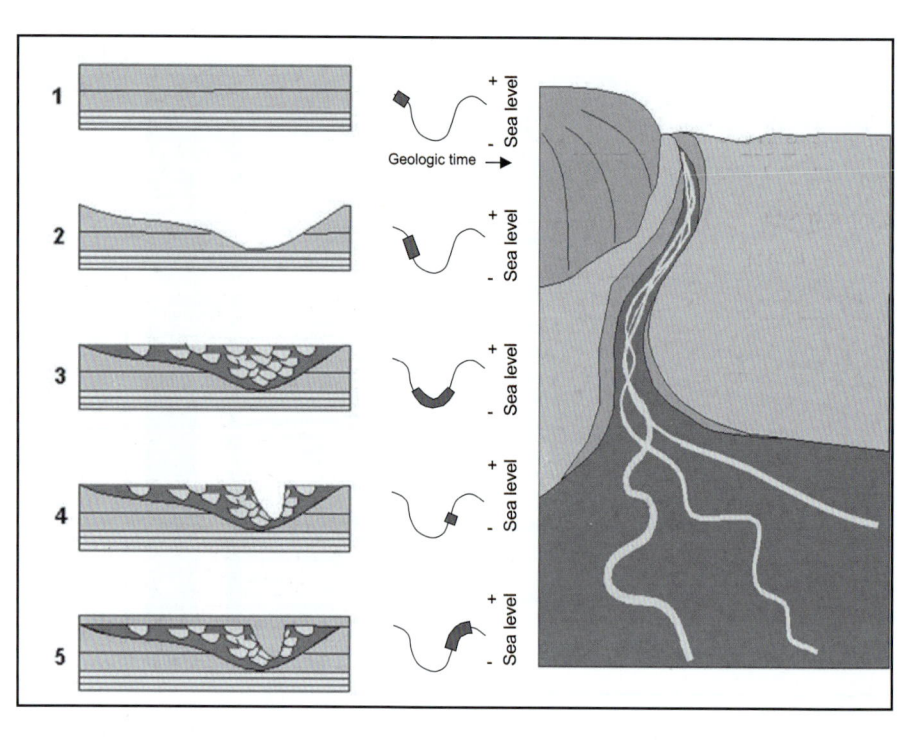

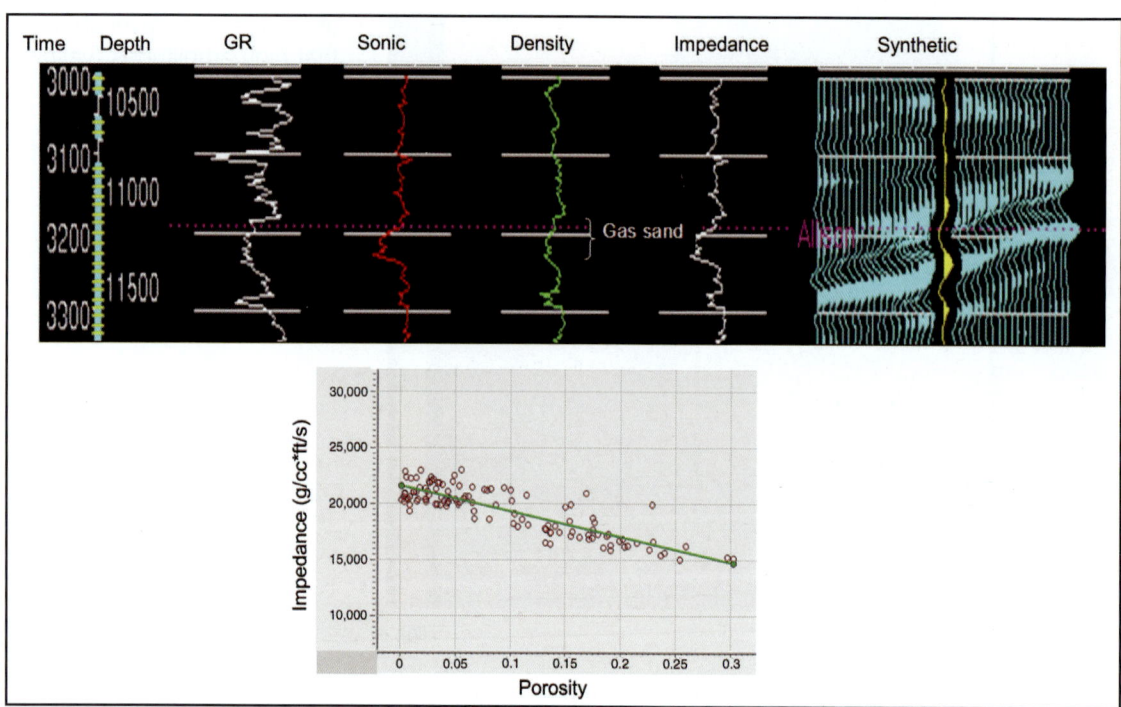

**Figure 6.** A synthetic seismic trace (yellow) is compared with the seismic data (blue) in the rightmost column. This is a good correlation that confirms from the other logs that gas sand is the cause of the high seismic-amplitude response. When porosity is plotted against impedance, the trend shows lower impedance at higher porosity. Lower porosities will respond with even higher amplitudes in this case.

Table 1. Reservoir parameters of the discovery well.

| Depth | 11,125 ft |
|---|---|
| Gross thickness | 171 ft |
| Net thickness | 100 ft |
| Porosity | 15–30% (average = 26%) |
| Permeability | 50–500 millidarcys (md) (192 average) |
| Area | 1.21e$^9$ ft$^2$ (27,778 acres @ 43,560 ft$^2$/acre) |
| Fluid | Retrograde gas tested at 32 MMcf/d |
| Pressure | 5990 psia |
| Average net porosity feet | 3.5 (because of stratigraphic variation) |
| Gas initially in place (GIIP) | 385 Bcf |
| Recoverable | 70% = 270 Bcf |

The reservoir summary is shown in Table 1. The reservoir is not huge, but if it is developed correctly, the operator should make money. The trick will be to keep down the costs of delineation and development, or the majority of future revenue will have been spent gathering subsurface information and/or getting the gas to market.

## Description of your decision responsibilities

*Your job is to select the delineation locations based primarily on the structure and amplitude maps.*

You can select to drill two to four delineation wells. The objective is to prove reservoir area and volume so that the project risk is acceptable for a final decision to develop the field, including the choice of facilities size. Right now, the only information you have is from a gas-filled well and from amplitude and structure maps that you might not trust completely. It is important to find out if the structural mapping is approximately correct, where the gas-water contact is (how full the trap is), and whether the reservoir is compartmentalized by faults or by stratigraphic barriers.

In general, you must determine the volume and extent of the reservoir so the right size of production facilities can be correctly designed and installed. If the facilities are too large and the accumulation too small, you will waste considerable money. If the facilities are too small, you will delay production and the stockholders will be poorly served. Therefore, the delineation program is an important part of the project.

Delineation wells are designed to determine location and volume of the reservoir, and you will drill dry holes when you drill outside the reservoir. Finding the water contact(s) is very important, and you might hit it just right. If you do, unfortunately, the well still would not be useful because it would produce so much water. But if you drill the delineation well gas full, it might be changed from a delineation well to a producer, saving you from drilling an additional well later. However, don't be surprised if all your delineation wells are plugged and abandoned — their purpose is to supply information and not to provide take points for reservoir production.

## Reservoir engineering and information about cost

The engineering department just has finished interpreting the flow tests taken in the discovery well, and they now have a good idea what the field will produce. A production profile for a projected three-well development scenario (Table 2) shows that gas is produced in only seven years. A market is available for the gas, and you look at the total project economics. At that point, the exploration team has sunk $84 million into the project, and you will need to cover that in profits. You also will need to cover the costs of the delineation program plus the drilling of development wells and installation of production facilities. All in all, you guess you might have $580 million to account for before you earn any revenue (Table 3). You need to get the project on stream very soon, because the stockholders want the discovery developed as soon as possible.

You have found an available rig and are responsible for site selection of the locations of two to four delineation well sites. Each delineation well will cost $30 million and will take four months to drill, so you have about a year to find out the real size of the discovery.

## Geophysical opportunity

At that point, the geophysics team tells you there is a very good chance that reprocessing the data with 3D prestack depth migration (3D PSDM) will obtain a much clearer subsalt amplitude map than the 3D poststack processing you are working with. It will take eight months to apply PSDM and will cost about $2 million, but the geophysics team thinks PSDM will help the delineation program by showing where to locate wells in the right places and thus determining the reservoir extent with fewer wells.

Do you want to do this or do you need to go ahead and drill now? You must decide. You also ask the team about the technology of acquiring an

Table 2. Preliminary annual production profile.

| Pressure | Production (billion cubic feet) |
|---|---|
| 5350 | 65 |
| 4200 | 61 |
| 3400 | 43 |
| 2900 | 34 |
| 2450 | 32 |
| 2100 | 20 |
| 1900 | 15 |

Table 3. Investment profile.

| | |
|---|---|
| Sunk costs today | |
| Lease | $20 million |
| Exploration well | $40 million |
| Geology and geophysics (G & G) | $19 million |
| General and administrative (G & A) | $ 5 million |
| Total | **$84 million** |
| | |
| Delineation and development investment | |
| Well costs — four(?) @ $30 million | $120 million |
| Geophysical program? | $ 2 million |
| G & A | $ 14 million |
| Development and facilities costs | $360 million |
| Total further project costs | **$496 million** |
| | |
| Earnings potential | |
| Revenue $3.33/mcf x 270 Bcf | $900 million |
| Less expenses | $580 million |
| Before-tax income | $320 million |
| After-tax (50%) income | **$160 million** |

additional 3D survey in a different orientation in hope of getting better data through acquisition. The team members agree that it might help if different azimuth data are acquired and added to the 3D PSDM program. They have done ray-tracing models and expect that benefits would be marginal. The cost of reacquisition and 3D PSDM together would be $12 million, and results will not be available for 18 months (Table 4). Do you want to do this? Do you hold up delineation drilling until you have those results?

## Decision table

The report you must make is represented in table form (Figure 7) for you to fill in. To help you in the location selection of delineation wells, a map of the amplitude with some key seismic lines is available to review (Figures 8 through 15). The lines are there to help you get a feel for the degree of variation in seismic data reflections from the reservoir. In general, higher reflections indicate better reservoir. However, remember that you are trying to find the entire size of the accumulation and not just the sweet spot, so you can't be afraid of drilling dry holes as long as they provide the information you need.

You also need to assess the seismic opportunity. Can you afford to wait for reprocessing or reacquisition of seismic data? You lose about 10% of the total value of the project for every year you delay getting gas to market. Better data might save you the cost and time of drilling one delineation well. Can you afford the seismic upgrade? Without doing an exhaustive decision analysis on the value of information, what's your opinion?

## Results

Decide whether and where to drill delineation wells. Also decide whether you are going to invest in an upgrade in seismic imaging. When these decisions are made, you can refer to the results of this project, as shown in Appendix A. There you can follow the delineation and development of the reservoir as it unfolded during a 10-year period.

Table 4. Seismic investment opportunity.

| | |
|---|---|
| Delineation well<br>○ includes conversion to producer<br>○ takes four months to drill each | $30 million |
| Seismic reprocessing<br>○ prestack depth migration subsalt<br>○ takes eight months to perform | $ 2 million |
| Seismic reacquisition and processing<br>○ change orientation for added imaging<br>○ combine and do prestack depth migration<br>○ takes 18 months for acquisition and depth imaging | $12 million |

Choose locations for two to four additional wells for delineation purposes (write down line and trace locations)

- More development wells can be drilled later.
- Is the reservoir full?
- Where is the water contact?
- Is the reservoir compartmentalized?
- Remember that you will drill producing wells later, after the delineation program.

| Well | Line | Trace |
|---|---|---|
| A | 3055 | 1470 |
| B | 3200 | 1506 |
| C | 3250 | 1490 |
| D | 3200 | 1500 |

Consider options for geophysical program

- Use 3D PSDM reprocessing.
- Delay drilling until after reprocessing.
- Choose reacquisition.
- Delay the delineation drilling program until after geophysics has been used.

|  | Yes | No |
|---|---|---|
| Use 3D PSDM reprocessing. |  | ✓ |
| Delay drilling until after reprocessing. |  | ✓ |
| Choose reacquisition. |  | ✓ |
| Delay drilling until after reacquisition. |  | ✓ |

**Figure 7.** In this student exercise, fill out the blanks in the tables with your choices to propose a delineation program.

**Figure 8.** In this poststack migrated amplitude map, use the line and trace numbers to reference the position of where you choose to drill delineation wells. The dashed red line is a structural contour and may represent the maximum gas/water contact.

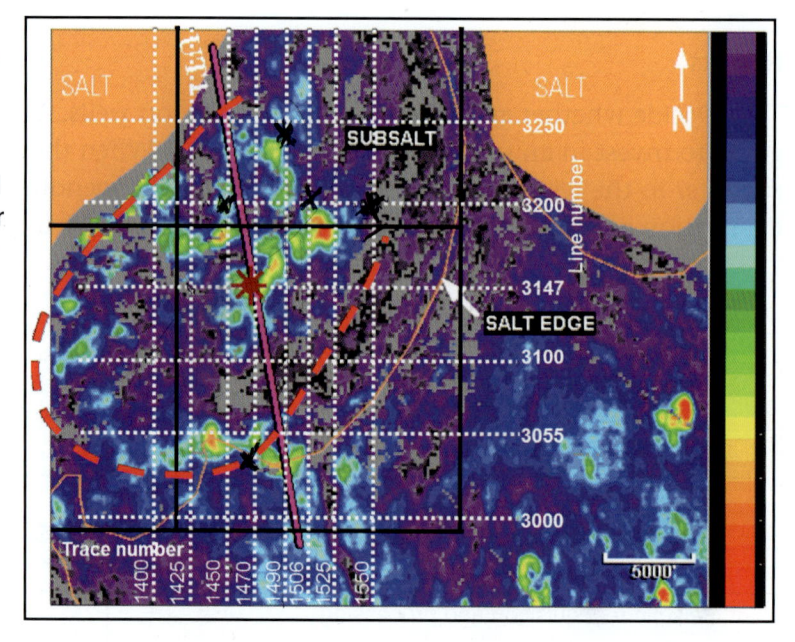

*Delineation Problem*

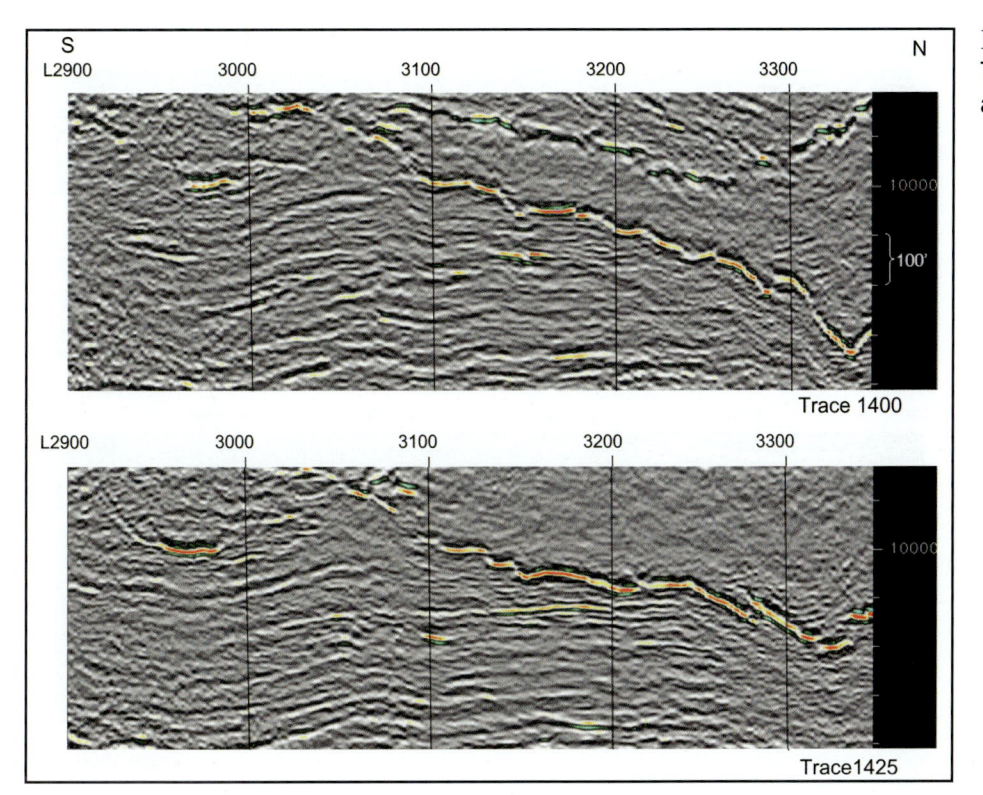

**Figure 9.** Traces 1400 and 1425.

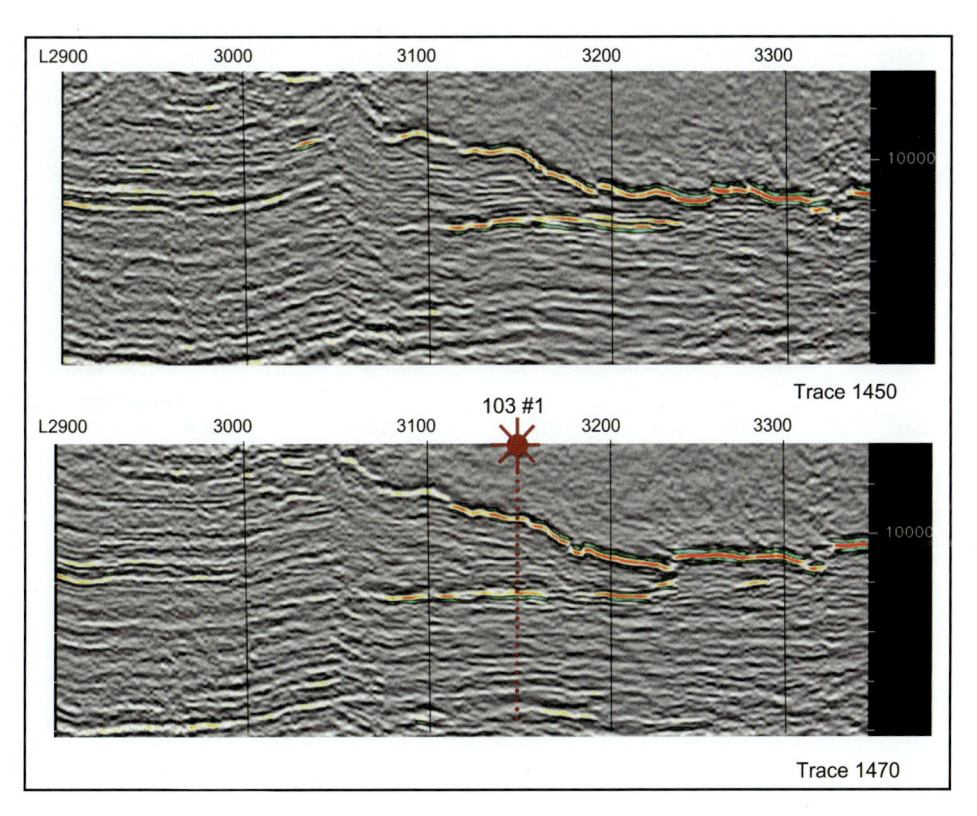

**Figure 10.** Traces 1450 and 1470.

*Distinguished Instructor Short Course* • 37

*Reservoir Geophysics: Applications*

**Figure 11.** Traces 1490 and 1506.

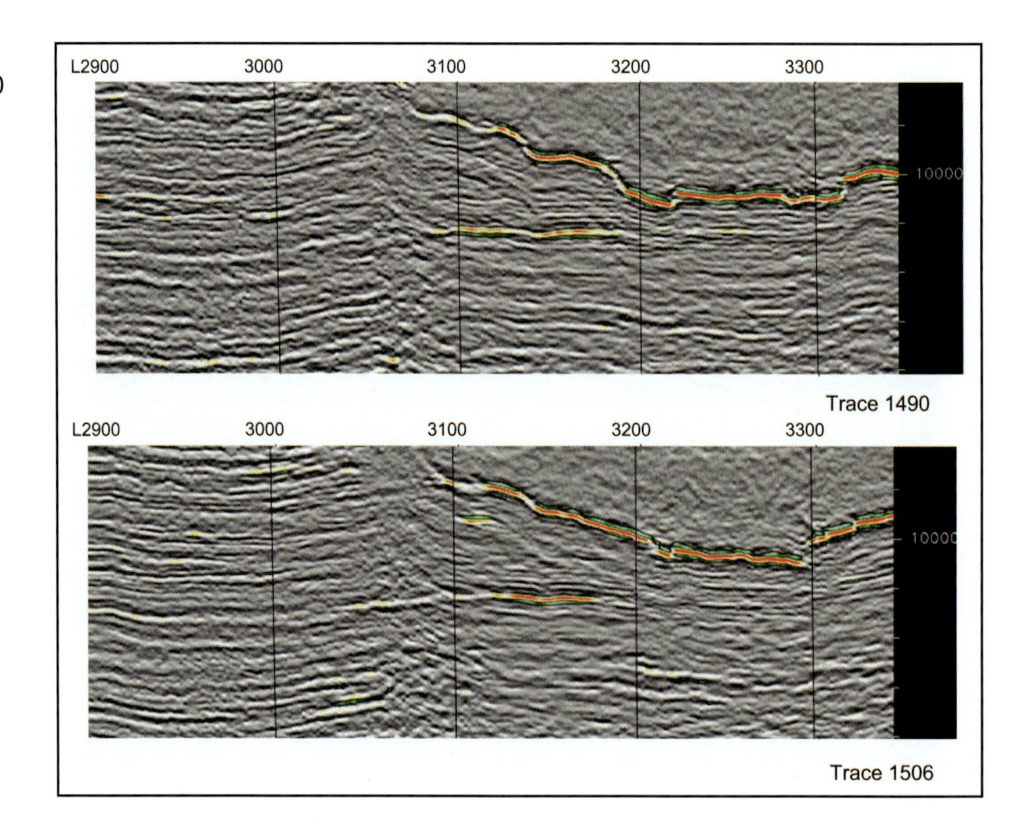

**Figure 12.** Traces 1525 and 1550.

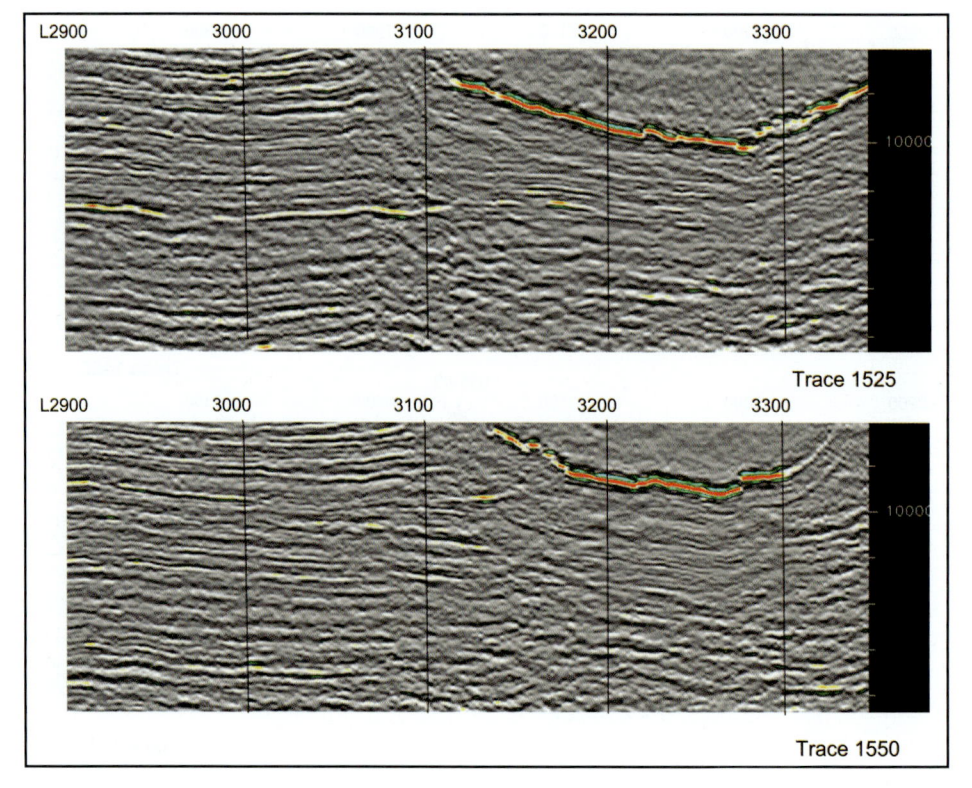

*Delineation Problem*

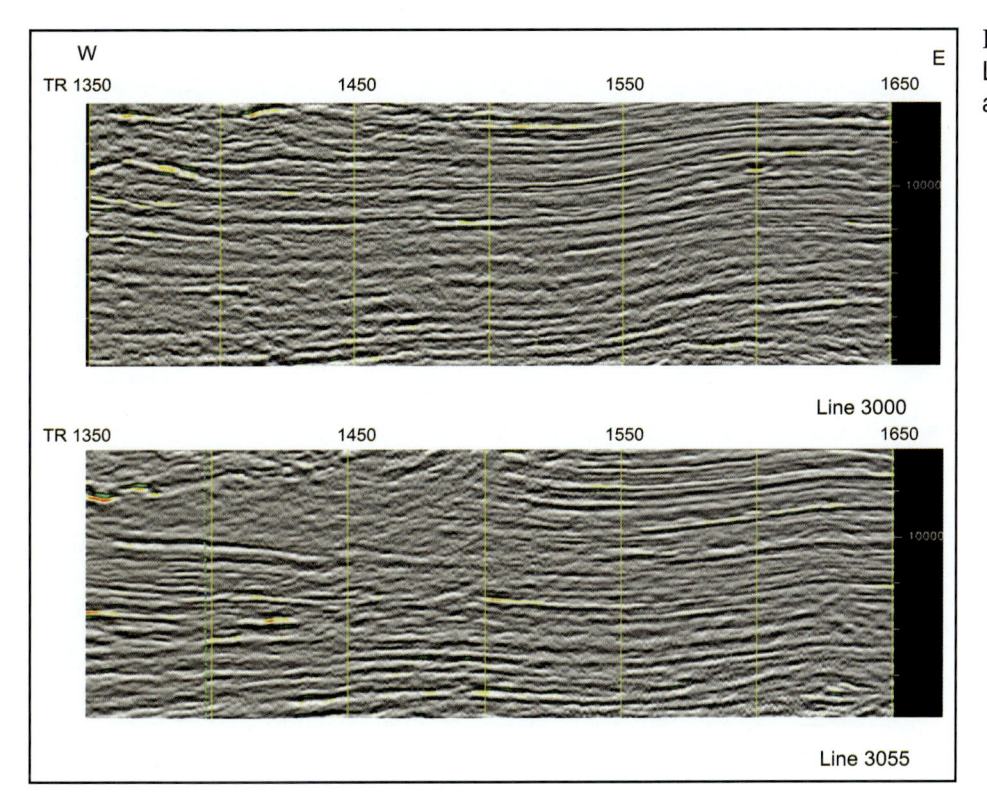

**Figure 13.** Lines 3000 and 3055.

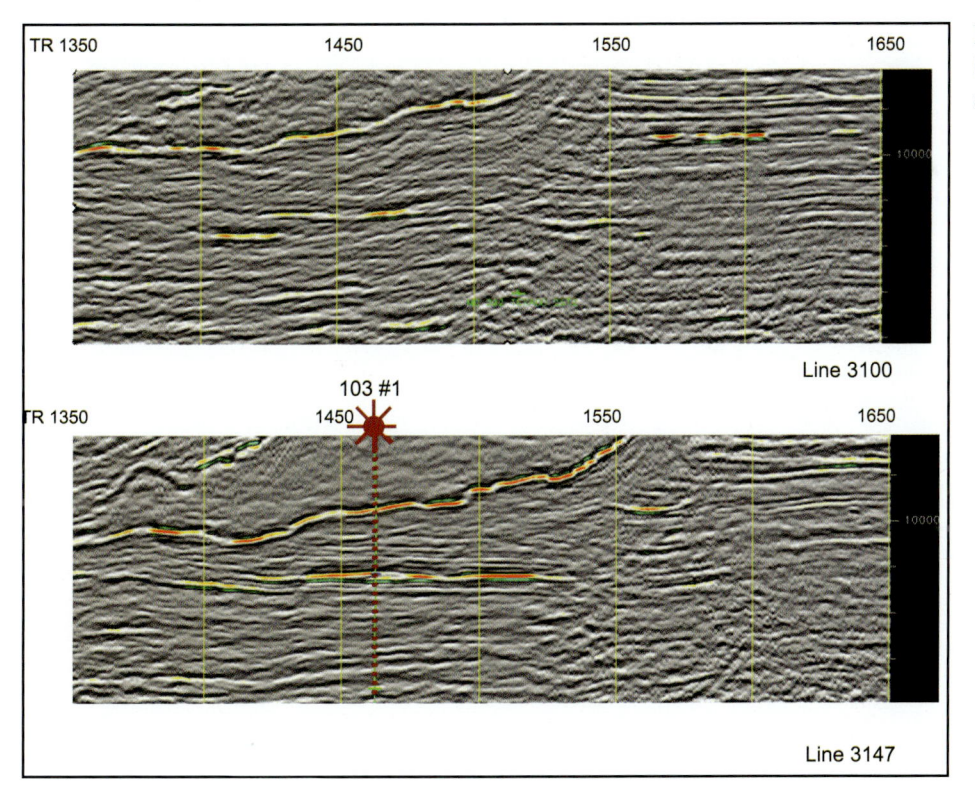

**Figure 14.** Lines 3100 and 3147.

**Figure 15.** Lines 3200 and 3250.

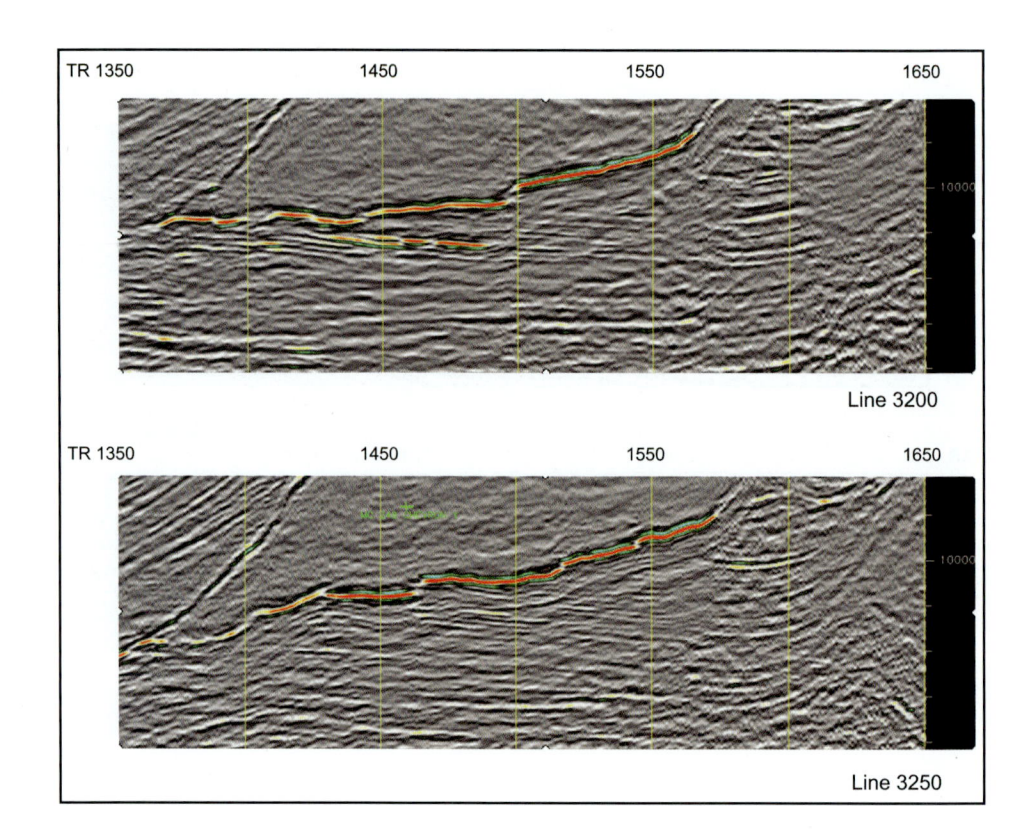

# Chapter 4  Development

## Economic drivers

The aim of the development phase in hydrocarbon field management is to establish an economic, sustained flow to market. To achieve that, several factors need to be balanced. Those include the number, location, and spacing of wells; the choice of fluid-handling facilities for water, oil, and gas; expected flow rates for individual wells; and the plan for recovery over the total life of the field.

During development, significant investments are made in drilling, completion, and facilities (Figure 1). To recover investments in a suitable time, initial production rates often are designed to be as high as facilities will allow, and production is designed to be sustained as long as possible. These variables will be balanced by the need to recover as much of the hydrocarbons as is reasonable under economic conditions, including secondary-recovery methods such as water flooding or steam flooding.

Development strategies are very dependent on subsurface geology. An example of that dependence for clastic reservoirs shows that total recovery for clastic rocks of some depositional environments can be as much as three times that of others because of the effect of lithology and layering of reservoir sands (Figure 2). To produce the field effectively, each development strategy requires a plan for drilling wells in specific locations, depending on the drainage area and productivity of each well plus the estimate of type of reservoir drive during production. Common drive mechanisms are described as bottom water, edge water, gas expansion, primary pressure, and reservoir compaction.

The project's prior delineation program was designed to gather information sufficient to answer those questions so the development program could move forward. The next challenge for subsurface geoscientists is to site specific well locations and to execute the optimal development program.

## Geophysics and reservoir characterization

During development, it is hoped that very few surprises will be encountered pertaining to presence and properties of the reservoir. Three-dimensional seismic data have proved to be a critical tool to aid in predicting the reservoir beyond the limits of well control. Seismic information is integrated with well data to provide a structural and stratigraphic model of the reservoir with best estimates of fluid location and type, and that information is used to site development wells. The integration of seismic and well data to estimate the characteristics of the reservoir is called reservoir characterization (Figure 3). Although it is possible to perform reservoir characterization without seismic data, those results are nearly always inferior.

Once the structure and the fluid contacts are established, the mapping of stratigraphic variations and fault boundaries of the reservoir helps to determine compartmentalization and differences in well deliverability. Successfully predicting those changes is the key to optimizing the development of the reservoir so hydrocarbons can be produced with the minimum number of wells. Because geology differs horizontally as well as verti-

**Figure 1.** Development drilling and facilities installation of the oil and gas project are the activities which require the greatest investment. The example shown is from an offshore project on the continental shelf. Onshore projects might not necessarily have such a high percentage of expenditures in production facilities.

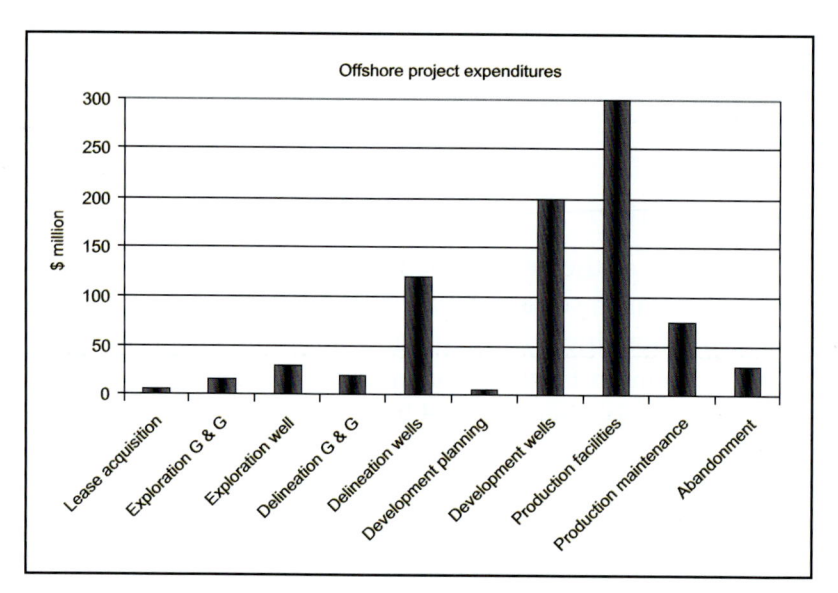

**Figure 2.** Data compiled from world records of recovery efficiency from clastic reservoirs in developed oil fields show wide variations from different geologic environments. Reservoir drainage is dependent on geology, and projects with poorer recovery require the drilling of more development wells (Larue and Yue, 2003).

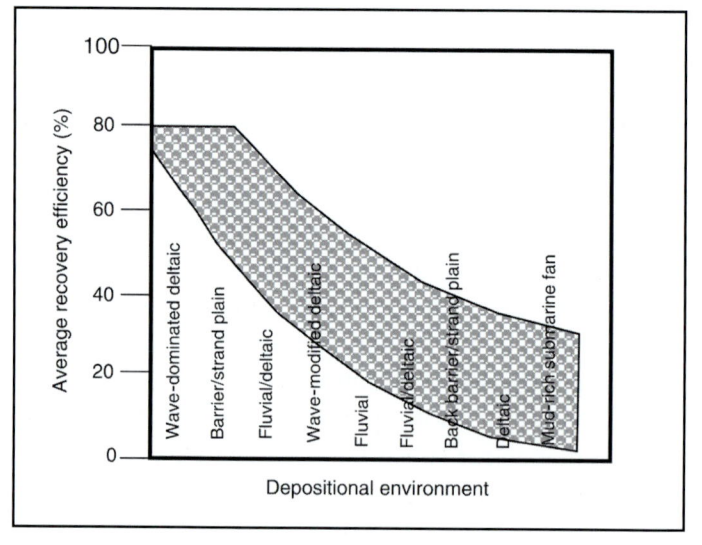

**Figure 3.** Reservoir characterization normally relies heavily on the use of seismic data. Wells and seismic attributes are combined in geostatistical models to produce fieldwide realizations of facies, geobodies, and porosity. The example shown suggests how complicated that process can become in the evaluation of fluvial clastic reservoirs (Liu et al., 2004).

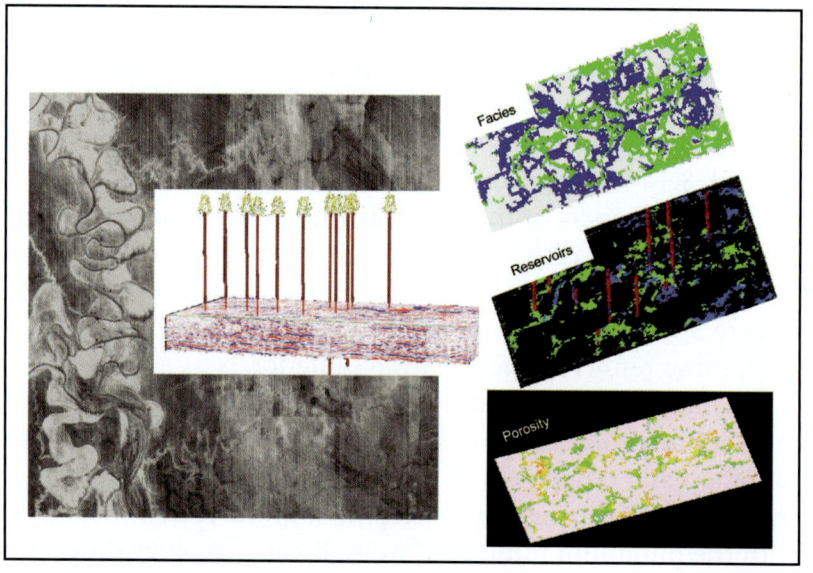

cally, local reservoir information from a well often does not successfully predict conditions in the rest of the reservoir. Changes in seismic patterns show important geologic features such as faults, unconformities, pinch-outs, and stratal patterns. In some cases, seismic data directly show reservoir fluids (Figure 4). The use of this geophysical information is very valuable in reservoir development, and it is important to find the right seismic attributes to provide the best information about the reservoir.

## Seismic response from reservoirs

Seismic data can be a powerful tool in describing the presence and quality of reservoirs. Three-dimensional analysis of seismic data patterns can be employed to fit the reservoir into the sequence-stratigraphic framework. Horizon slices of 3D seismic data sometimes can be used to determine locations of depositional facies (Figure 5).

Analysis of seismic data would be an easier task if every reservoir responded to seismic data in the same manner, but the acoustic parameters of rocks vary significantly. Modeling the entire expected range of rock properties that can be encountered is important. The range is almost certainly greater than that encountered in delineation wells, and it must be accounted for to correctly interpret the seismic data response between wells. A large variation is seen in seismic response for even one reservoir because acoustic response is a function of bed thickness, overburden and substrate lithology, grain sorting, pore shape (Figure 6), diagenesis, grain-grain contacts, fractures, pressure, and fluids.

The influence of rock texture on seismic response is very important and must be taken into consideration for predicting reservoir properties. For instance, in clastic rocks, variation in the ratio of clay and sand grains shows that velocity is not a linear function of the percent of shale (Figure 7). That is one of many factors that makes prediction of reservoir properties from seismic data a nonunique solution.

Seismic modeling for reservoir attributes often is done by taking a representative well and modifying rock and fluid parameters so that time and amplitude changes on a

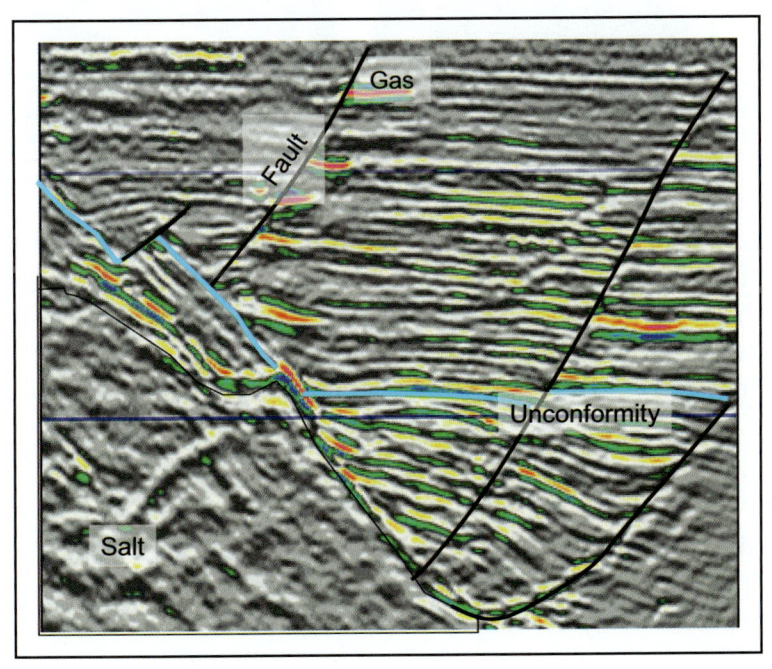

**Figure 4.** Seismic data can provide rich information on lateral differences in subsurface geology, including faults, unconformities, stratigraphic changes, and fluid content.

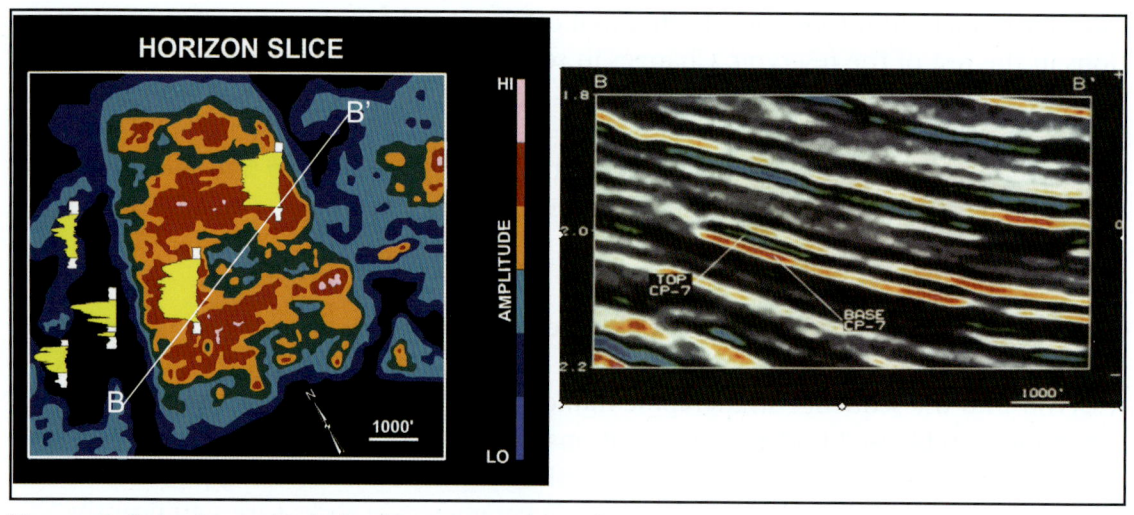

**Figure 5.** Seismic amplitude is often a powerful attribute. The horizon slice of the CP-7 sand shown here illustrates the relationship of porous channel sands to high-amplitude seismic response (Abriel et al., 1991).

**Figure 6.** Seismic response is a function of many variables, and it is not likely that any one can be isolated easily. For example, note how different the grain-supported pore geometries are in one carbonate rock than the dissolution pores are in another. Because of the wide range of geology in the subsurface, predicting absolute amplitude values often is not as useful as understanding spatial patterns of relative amplitude changes (Eberli et al., 2003).

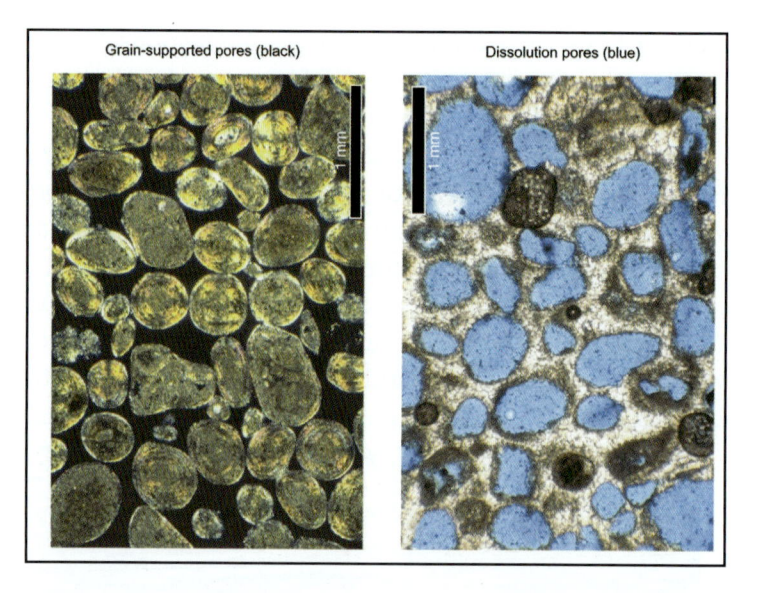

seismic trace are considered. A danger to be avoided is to hold too strong a belief in the first conceptual model of the reservoir. This limits consideration of wider ranges of geologic outcomes in development drilling. At times, modeling also includes 2D sections of simple modifications in rock character such as reservoir thinning. Sometimes a geologic model also is generated with changes in rock and fluid characteristics. When the model matches seismic lines in 2D and 3D, confidence in the understanding of the reservoir properties is raised.

## Seismic attributes

How then does one find the right seismic attribute for reservoir prediction? Certainly, it is critical to establish the well response and the seismic tie. By scanning a wide range of geologic possibilities represented by 1D synthetic seismograms against the range

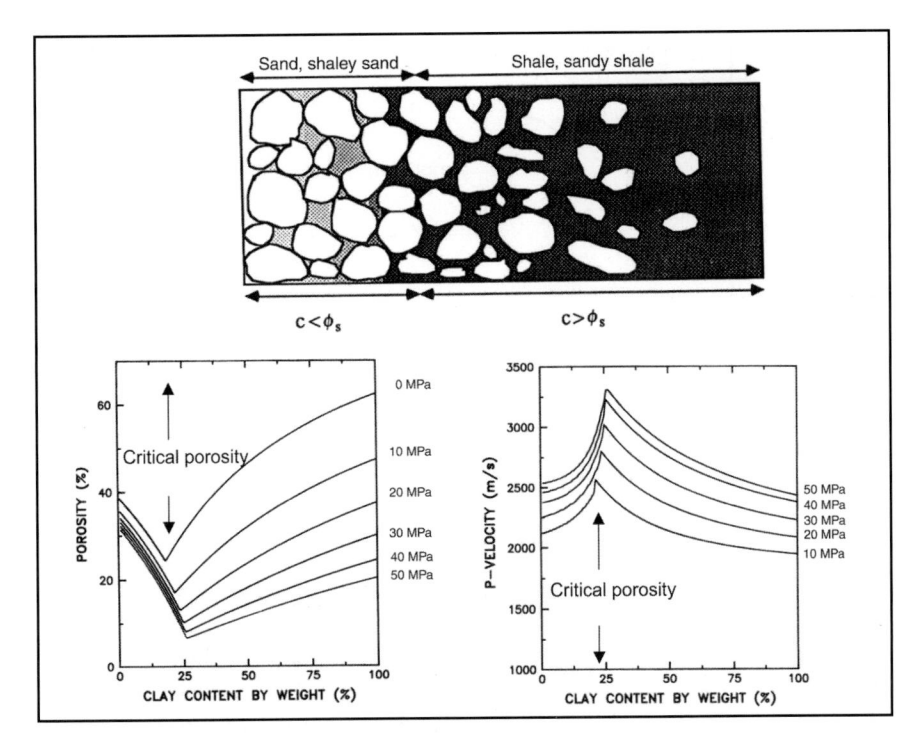

**Figure 7.** Velocity of seismic data is not a linear function of lithology in grain and clay depositional systems. Note the highest density and velocity peak at the point of critical porosity, where clay has filled the pores between grains (Marion et al., 1992).

of seismic attributes, prime-candidate attributes emerge. World experience shows that some attributes are almost always significant, and it is reasonable to try those early in the test. They include velocity, amplitude, frequency, shape, and coherence (Figure 8).

Seismic data in trace form can be recast in attribute form for better viewing and analysis. As an example, the traces can be transformed to show total reflection strength, which provides a more intuitive interpretation than do traces alone (Figure 9).

Another attribute especially useful for recognition of stratigraphic changes is frequency or spectral decomposition of traces. When shown in horizon-slice view, different stratigraphic elements of the reservoir are seen more easily (Figure 10).

Classification of seismic traces on the basis of overall shape is also helpful. A case history by Poupon et al. (2004) shows how classification of seismic data on the basis of waveform shape is used to identify depositional facies of the reservoir, which then can be used for planning development drilling and production drainage (Figure 11).

Volume attributes are those that relate traces to one another and are computed over a 3D cube. A commonly derived attribute is 3D coherence, which is a measure of how consistent the traces are. When coherent parts are seen as connected, the image often allows for fine detail in determining faulting, fracturing, and fine-scale stratigraphy. A case history by Skirius et al. (1999) shows fine-scale faults determined from 3D coherency processing of a carbonate reservoir in Canada (Figure 12).

As can be seen from the above examples, an important step in attribute work is to view seismic-attribute candidates in map form. Viable attribute candidates reveal geology and help determine potential stratigraphic models to be considered in development planning. At that point, it becomes helpful to minimize the number of attributes used in reservoir prediction so that only important contributors remain. A good way to help employ attributes without imposing a user bias is to use neural-network prediction.

*Reservoir Geophysics: Applications*

**Figure 8.** A single seismic trace can provide three of the more common attributes of seismic response — amplitude, frequency, and shape.

Common seismic-data attributes

- amplitude
- frequency
- shape
- coherence
- velocity

Amplitude

Period = 1/frequency

Amplitude

Reflection strength

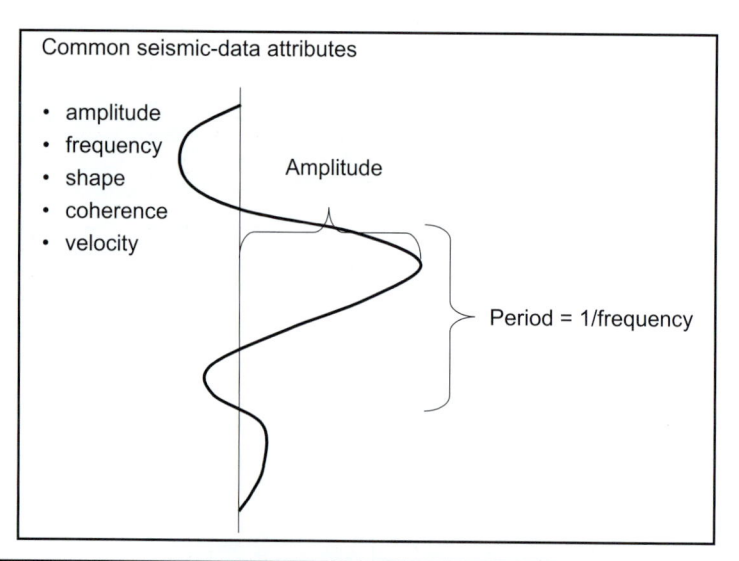

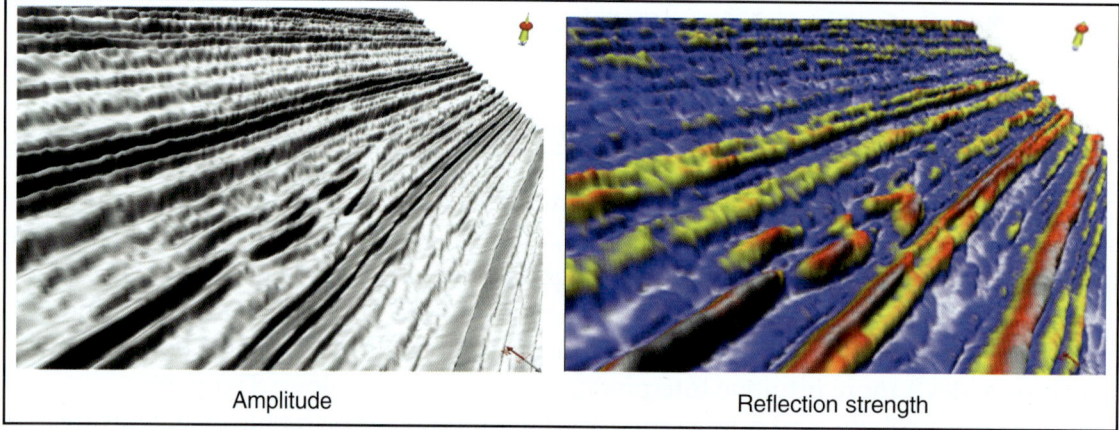

**Figure 9.** A shaded perspective display shows the difference between seismic traces and the attribute of reflection strength. Strength is an attribute calculated from the seismic trace. It is the algebraic sum of the kinetic energy (recorded real trace) and the computed potential energy (Lynch and Lines, 2004). See Sheriff (2002) for details on reflection strength.

Energy envelope

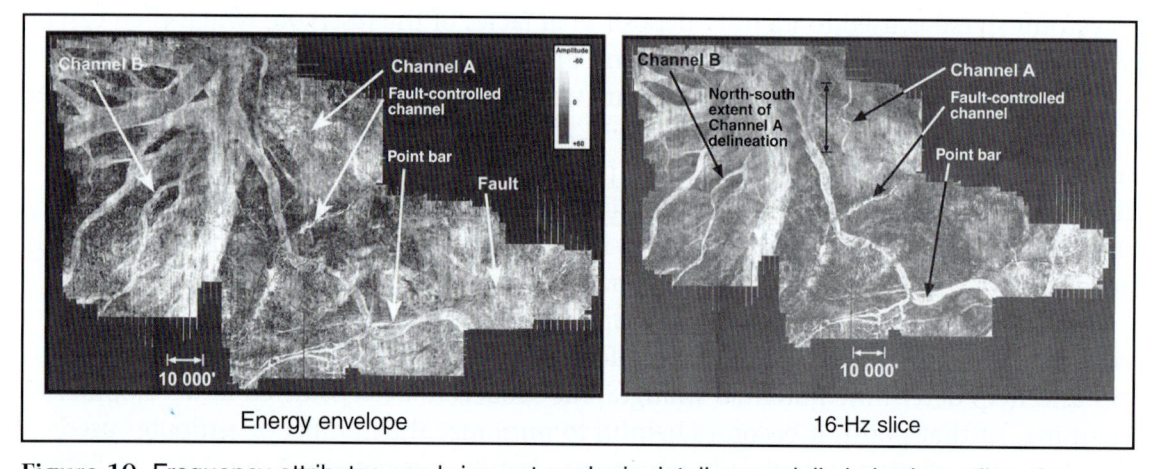

16-Hz slice

**Figure 10.** Frequency attributes can bring out geologic detail, especially in horizon-slice view. Here, the frequency slice at 16 Hz highlights channels and faults not seen as easily in the attribute of the energy envelope (similar to reflection strength) (Partyka et al., 1999).

46 • Society of Exploration Geophysicists / European Association of Geoscientists & Engineers

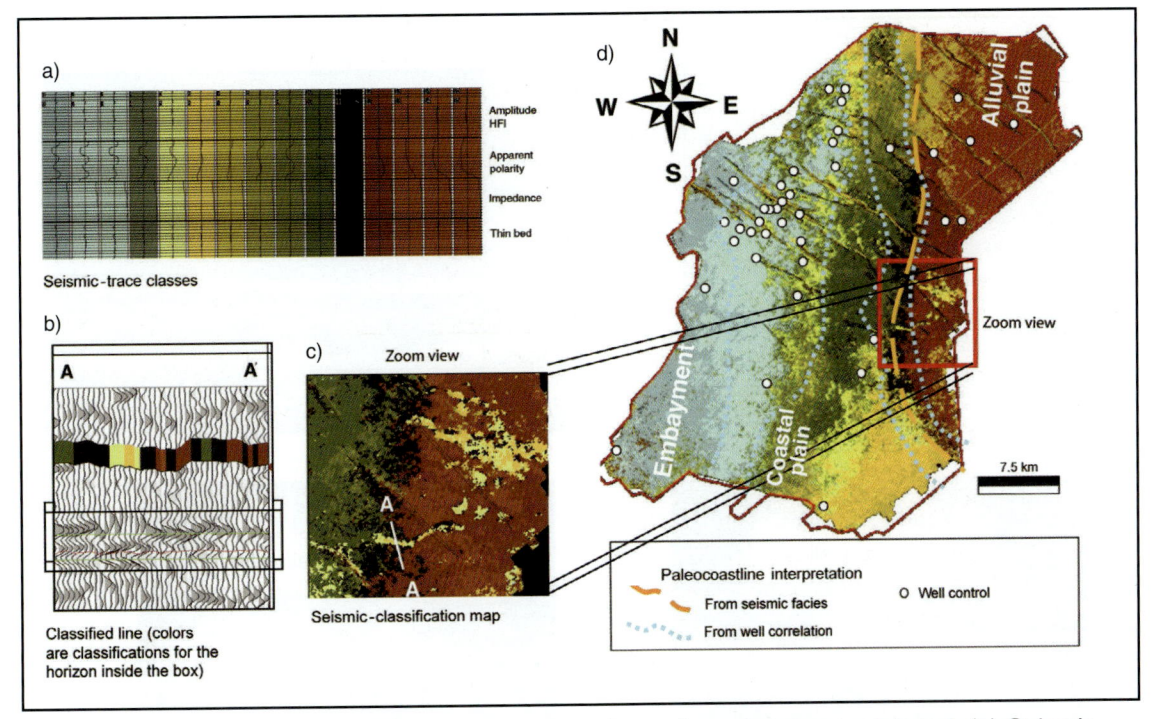

Figure 11. The shape attribute measures the waveform character over an interval. (a) Seismic traces are examined and sorted into a finite number of classes. (b) Classes are related to variations seen in seismic traces at the horizon of interest. When seen in (c) map form, the seismic classification can be a powerful tool for (d) stratigraphic interpretation (Poupon et al., 2004).

## Attributes workflow

A recommended workflow for attribute selection and use includes using well data and a conceptual geologic model along with seismic data (Figure 13). Using well data alone to generate expected attribute responses can lead to a wrong interpretation because well data are only a subsample of the geology in the reservoir. It is best to consider the entire geologic range (the larger conceptual geologic model) and to have enough model responses generated to cover the range of seismic responses in the data.

Independently, seismic data also are scanned for what you think are the right attributes. The results normally are reviewed by visual interpretation to see if they reveal the geology. That is confirmed by forward seismic-trace modeling before choosing the small number of attributes from which to predict the reservoir. When used in that way, attributes have been found to be a very powerful predictor of reservoir presence and quality.

## Inversion

If seismic attributes are effective, it is because they act as a proxy for the seismic data response of the reservoir. Another important representation of seismic data is to transform them directly to acoustic impedance, removing the wave nature of the data. That process, known as inversion, produces seismic traces that represent acoustic well logs as closely as possible (Figure 14). Inversion requires very detailed knowledge of the wavelet in the reservoir zone, an excellent measure of local rock velocity, and an under-

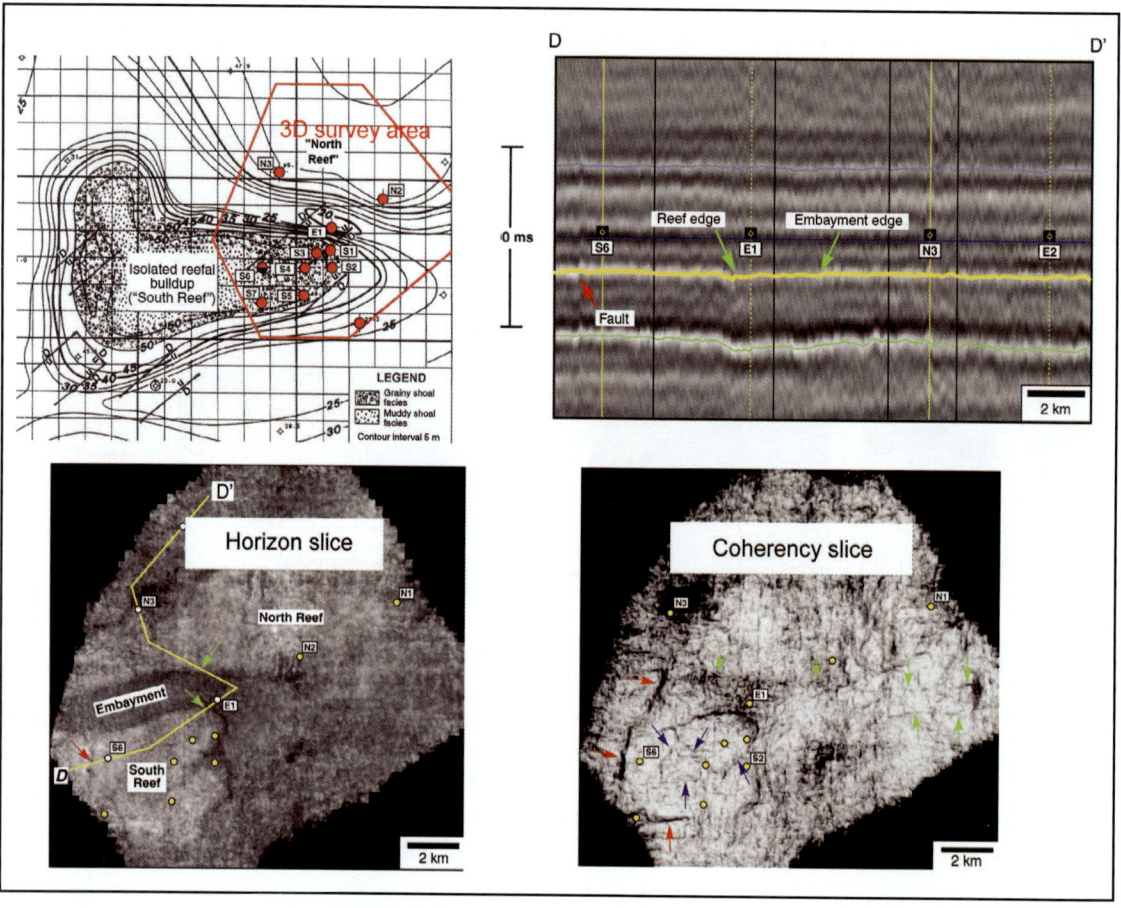

Figure 12. Volume seismic attributes relate the changes of the seismic variations in three dimensions. An example of that is coherency attribute. In this figure, the faults of the horizon outside and inside the carbonate reservoir are enhanced significantly (Skirius et al., 1999).

Figure 13. In a recommended workflow for calculation and selection of seismic attributes, seismic data and well data are analyzed separately at first. Once the independent information they contain is understood, they can be used together for predicting reservoir properties.

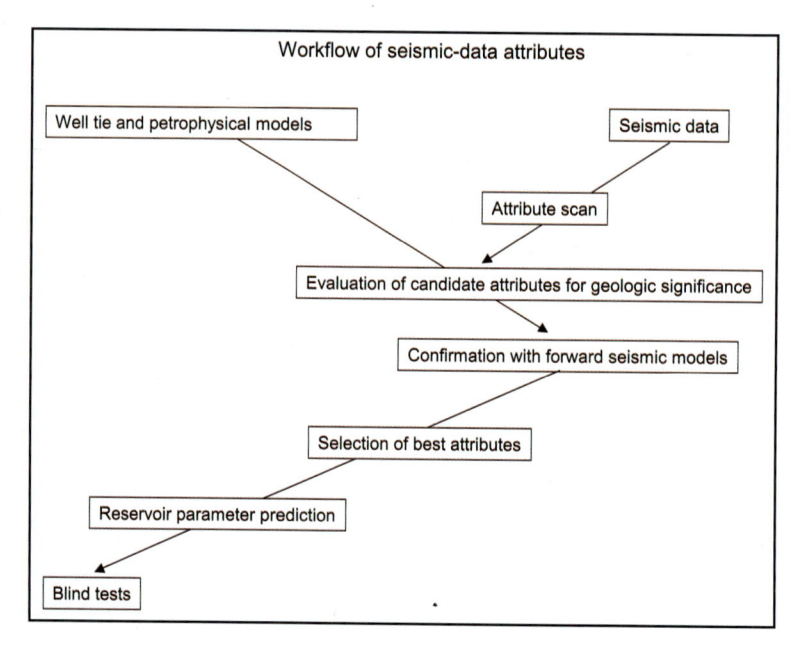

standing of the geologic model used to guide results. Inversion of seismic traces produces an image that is more intuitive for stratigraphic interpretation (Figure 15). Under the right conditions, inversion shows location, thickness, and stratigraphic variation of the reservoir directly, but it is limited by the resolution of the seismic data because of transmission frequency loss (Figure 16).

## Inversion case example

For each project in development, an important contribution from seismic data can be made by using the right attributes. Examples of applications to many projects

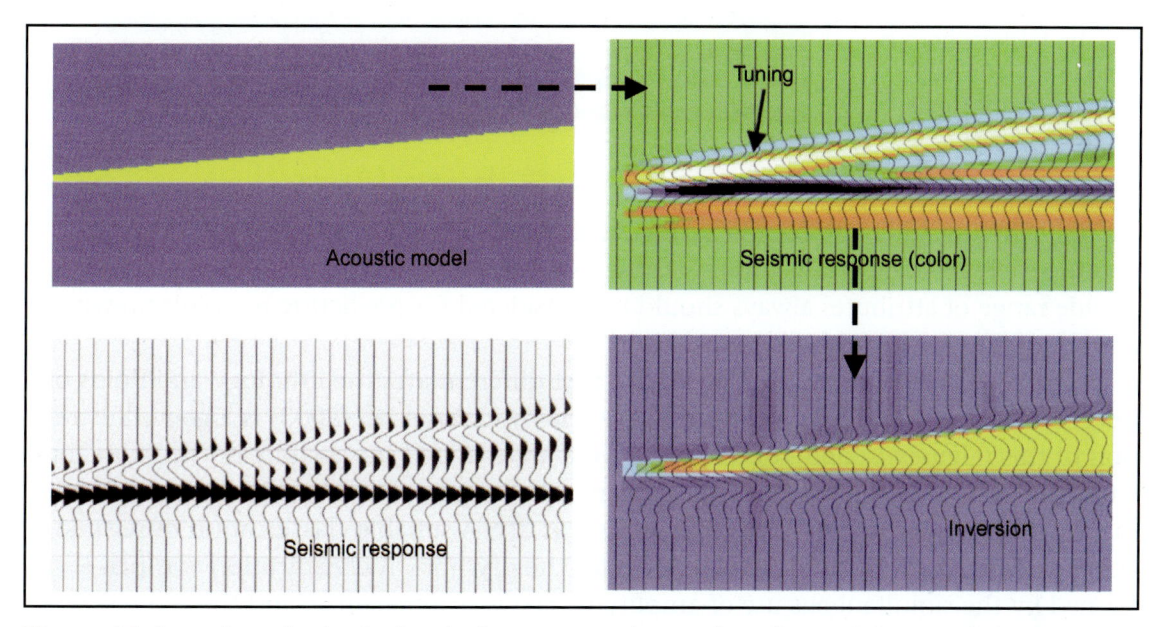

Figure 14. Inversion of seismic data is the process of removing effects of the wavefield so the response most nearly resembles acoustic logs. That requires exceptionally good knowledge of the wavelet in the data as well as precise information on the velocity of the zone of interest. Here, the inverted data closely resemble the starting model, as they should (Latimer et al., 2000).

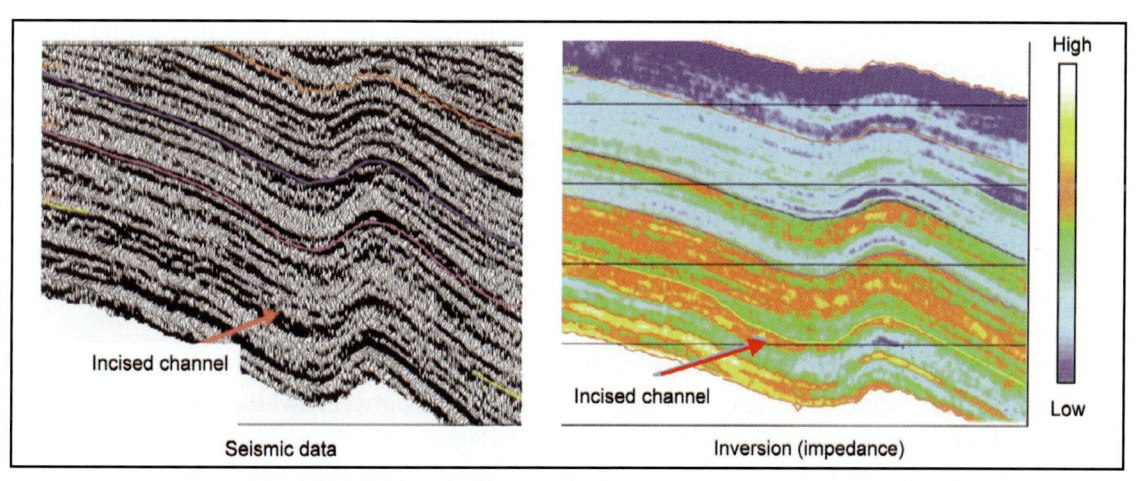

Figure 15. An example of the application of inversion shows several high-impedance depositional cycles, one of which is incised by a low-impedance channel (Latimer et al., 2000).

**Figure 16.** The inversion shown is targeted for carbonates at 10,600 ft. On the right, the sonic log of a dry hole is beside the synthetic inverted seismic traces. The zone of interest in the deep part of the section shows low impedance and high porosity in lighter colors. Note that the thickest porous zone in the gas well extends about 1 mile (Lindseth, 1979).

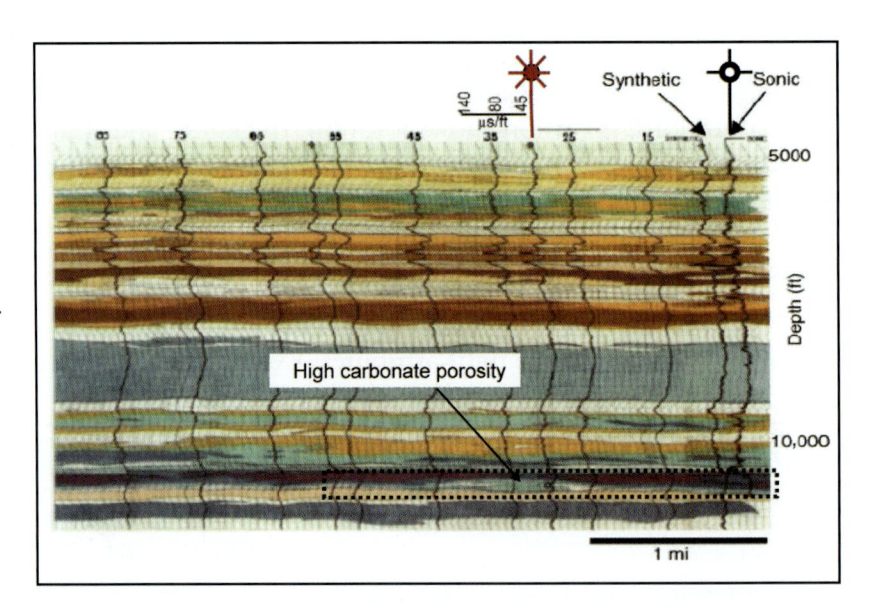

are published steadily and should be reviewed as suggestions of what to try. However, because each geologic area is unique, those examples should be used only as guidelines. A wide range of attributes always should be considered for predicting reservoir presence and properties for your project.

Often, a specific mixture of attributes best differentiates the lithology and fluids of reservoirs. An example from onshore Venezuela by Linari et al. (2003) is a good illustration of this (Figures 17 and 18). Three oil wells were drilled successfully on structure, but a fourth was nonproductive. Neither the structure map nor the horizon slice (amplitude map) could show why one of the four wells was dry. A suitable classification of seismic traces revealed a stratigraphic explanation for the dry hole. A classification was performed by the simultaneous use of amplitude, acoustic impedance, and the AVO attribute fluid factor. Interpretation of the results became a basis for further development of the reservoir.

## Reservoir flow model

When the development drilling plan is to be initiated, the best possible understanding of the reservoir is needed so the effort can be optimized. Although that understanding includes seismic data as an important component, reservoir-engineering estimates of fluid recovery will influence the plan even more. In an important field, reservoir engineers normally develop several reservoir-simulation models with differing well positions and reservoir properties to predict the production of the field (Figure 19). That is an important part of production forecasting and management of the reservoir. Often, the reservoir model is influenced strongly by seismic data with regard to compartmentalization and reservoir properties.

In reservoir forecasting, data derived from delineation wells are critical in building valid simulation models. However, reservoir information between wells is also important, and seismic data have a major role to play in that (Figure 20). Commonly, seismic data do not directly measure the reservoir property of interest, certainly not at high enough resolution. Seismic data are used more commonly to constrain reservoir models and provide boundaries on uncertainty.

## Development strategy

The important questions to answer during development are: (1) Where exactly should production wells be located? (2) What area will each well drain? (3) What areas are connected by pressure? The model of the subsurface needs to be of sufficient quality to answer those questions as accurately as possible. Although they are imperfect measurements for the responses to the questions being asked, seismic data have been proved to be a critical contributor to the reservoir prediction model.

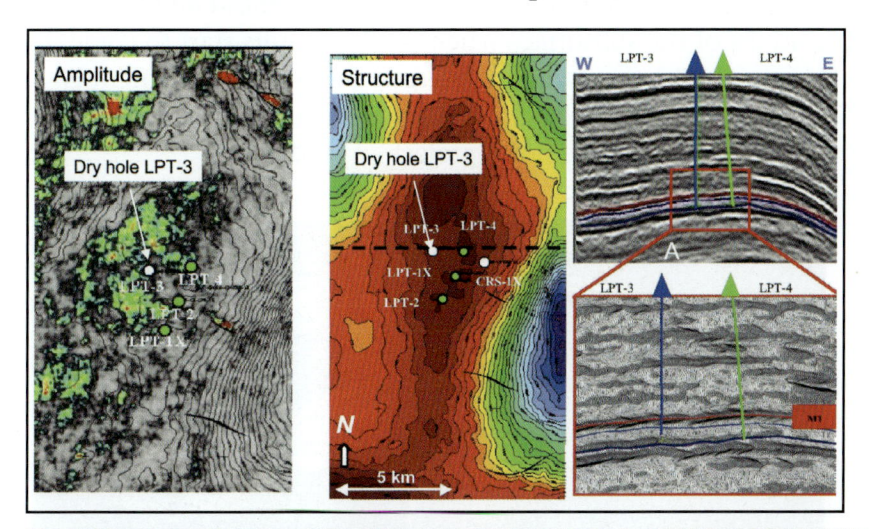

Figure 17. Some of the wells drilled on structure in this clastic sandstone reservoir are nonproductive. Seismic data were used to derive structural position and amplitude response. Neither of those attributes seems to explain the dry hole at LPT-3 (Linari et al., 2003).

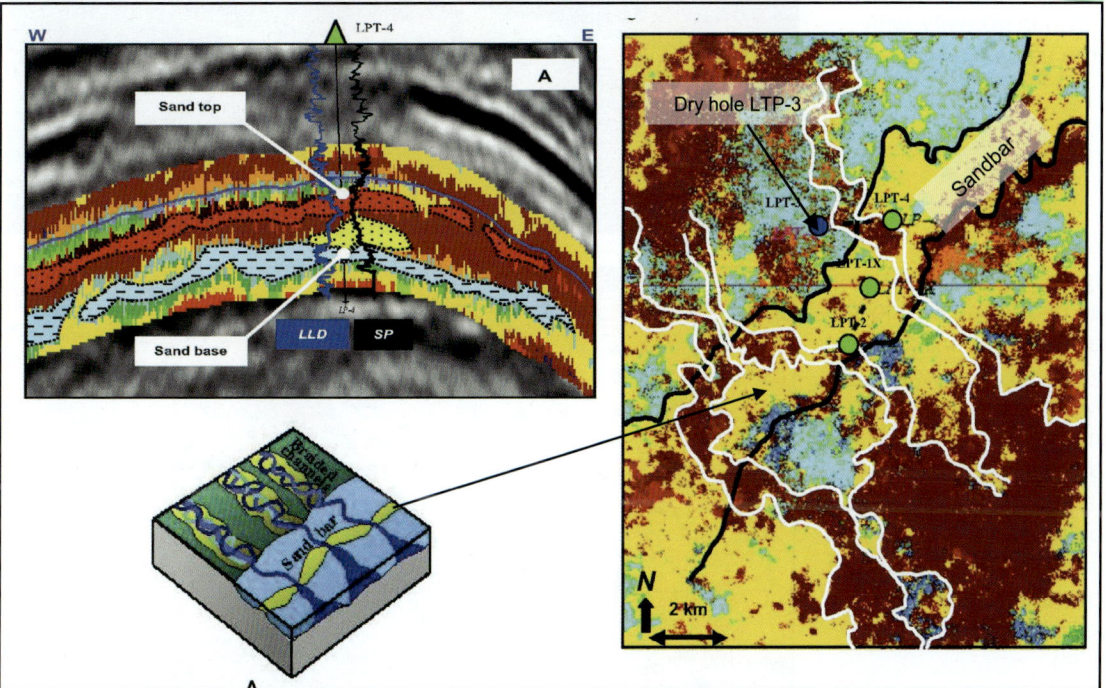

Figure 18. Through use of the combination of seismic attributes of amplitude, impedance, and AVO fluid factor, lithology of the reservoir can be discriminated. Colors represent different lithologies. The interpretation is that the dry hole was drilled outside the sandbar, which is the productive facies of this reservoir (Linari et al., 2003).

Once development drilling starts, it can be difficult to adapt to unexpected outcomes in the subsurface. Despite the best technology, surprises often are encountered in presence, quality, and connectivity of the reservoir. To help us adjust to new subsurface information, fluid tests, geochemical analysis, and well-flow tests can ensure connectivity of the reservoir and deliverability of wells. If changes in the development program are required and wells need to be relocated, the challenge is to upgrade the subsurface interpretation rapidly.

Therefore, during development, it is important to keep an open mind and be prepared for different interpretations of the subsurface when drilling results start to come in. The development phase is normally the time when the largest monetary investment is made and the largest volume of subsurface well information is obtained. That information is reviewed for optimizing the next phase of the project — production.

**Figure 19.** Extensive planning is required prior to development drilling. Reservoir engineers are likely to generate a flow model from well and seismic data to predict compartmentalization and drainage. Here, the reservoir-model attribute shown is porosity, with large values in red.

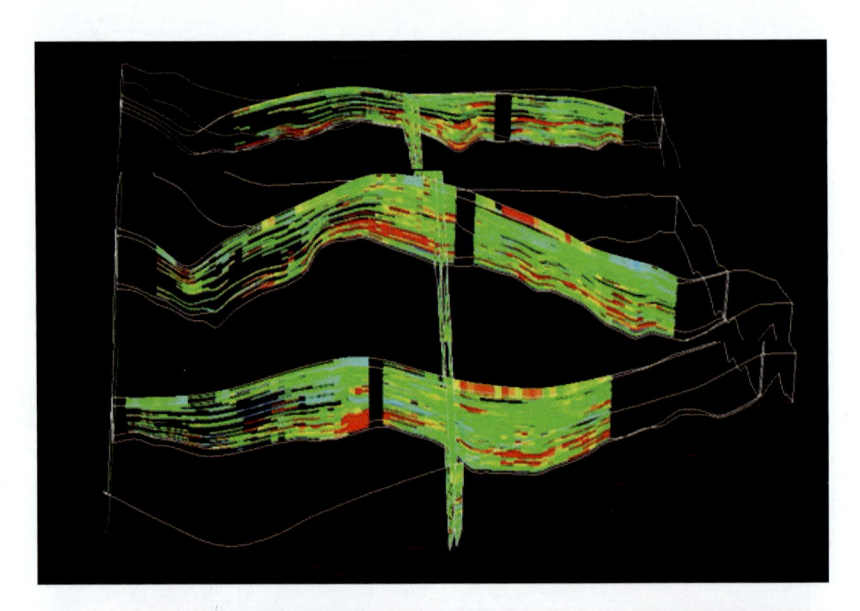

**Figure 20.** Seismic data can be a powerful tool in development drill planning. This horizon slice of inverted acoustic impedance shows fluvial sand compartments in red. A plan to develop the reservoir must predict how well sands are connected and how large an area each well will drain (Liu et al., 2004).

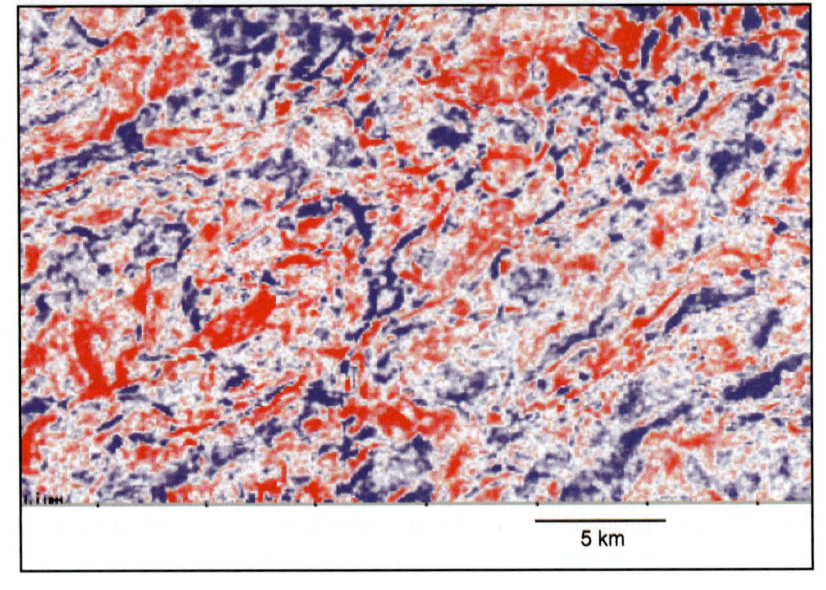

# Chapter 5   Production

## Economic drivers

Once the development phase of the field is complete, the field enters into production, and the hydrocarbons are sent to market. During that time, it is most important to ensure delivery, maximize production, and minimize operating expenses. Reservoir management requires the ability to (1) ensure flow throughout the production system, (2) manage gas movement caused by pressure changes, (3) manage the tendency of water to move to the well before oil, (4) account for bypassed reserves, and (5) maintain pressure for driving production.

After the intense activity required to find, delineate, and develop the field, it would be comforting to think that complexities of the subsurface are resolved and the field can simply make money. That is not the case, however, because the flow properties of fluids and the heterogeneity of rocks now begin to impact the project.

As production commences, the fluid dynamics of the reservoir become all-important. As hydrocarbons are produced, changes occur in the reservoir itself, including fluid movement in pore spaces, pressure reduction, gas going into or coming out of solution, and reservoir compaction. Some of those dynamic variations might occur quickly, whereas others might take years to have effect.

Undesirable production characteristics can occur because oil, water, and gas do not flow through rocks at the same rate. Depending on the characteristics of oil and reservoir formations, operating the field at high production rates can cause gas or water to reach the borehole earlier than necessary, significantly decreasing project value (Figures 1 and 2).

## Production geophysics

In geologically simple reservoirs, changes are monitored adequately at production wells and are predicted accurately from reservoir simulation models. However, in dealing with geologically complex reservoirs, appropriate monitoring becomes more important. Complex reservoirs can be monitored with downhole instruments that gather significant reservoir information such as pressure, temperature, and flow rates and that relay the information to production supervisors. Then that remotely acquired information can be used to manage some of the production elements for optimization — for example, to increase or decrease production flow from a specific interval in a specific well.

Under the right conditions, seismic data also can be used as a tool to monitor reservoirs during production. Where fluid type and saturation affect the seismic reflectivity of the reservoir, seismic data can be used to understand the production state of the reservoir (Figure 3). Fluid type and saturation are major factors in the understanding of fluid dynamics of the subsurface, and knowledge of these parameters is critical for good reservoir management.

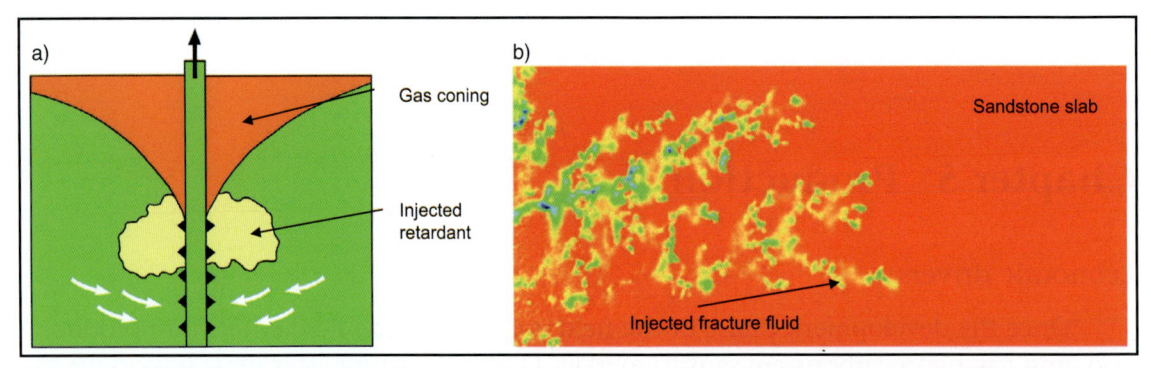

**Figure 1.** (a) During production, the gas cap might be produced faster than oil, resulting in coning from above. Once that occurs, gas preferentially moves through the affected pores, and oil flow is restricted, unless a chemical plug is used to restrict gas. (b) Fingering of gas through a sand slab 60 × 25 × 0.6 cm saturated with hydraulic-fracturing fluid is shown. Note the fingering of gas and the bypassed areas (Tidwell, 2007).

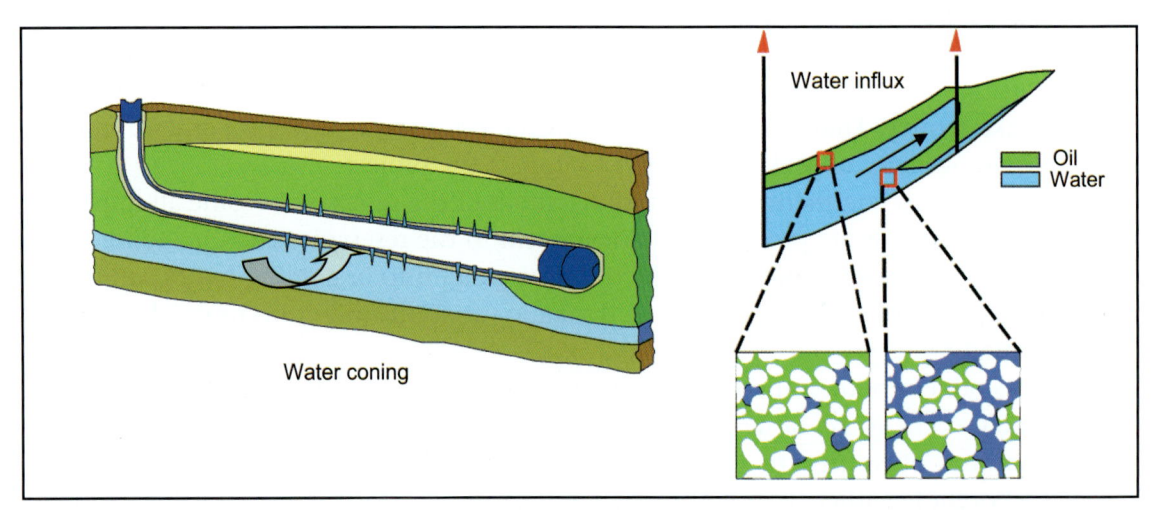

**Figure 2.** Water generally moves more easily in reservoirs than oil does, particularly in heavy-oil reservoirs. During production of fluids, water might reach the producing well through fingering or coning. Once that occurs, the well will produce a large percentage of water.

When the reservoir model is generated before production, characterization is completed after development drilling, and it constitutes the base case for reservoir simulation and production forecasting. As discussed in the section on development, seismic data can have a substantial impact on the initial reservoir model. That is because the data are direct observations of reservoir properties between wells and represent the static conditions of the reservoir. However, if seismic data are acquired after production begins, the effects of the fluid dynamics have to be taken into account in the interpretation of seismic observations (Figure 4).

Interpretation of data can be difficult and ambiguous if the reservoir is not behaving the way reservoir simulation had suggested. Compartmentalization by small faults and stratigraphic barriers can retard flow of oil and gas. Alternately, well-connected high-permeability zones can cause bypassing of parts of the reservoir or preferential channeling of aquifer waters. In both bases, the reservoir will be produced unevenly.

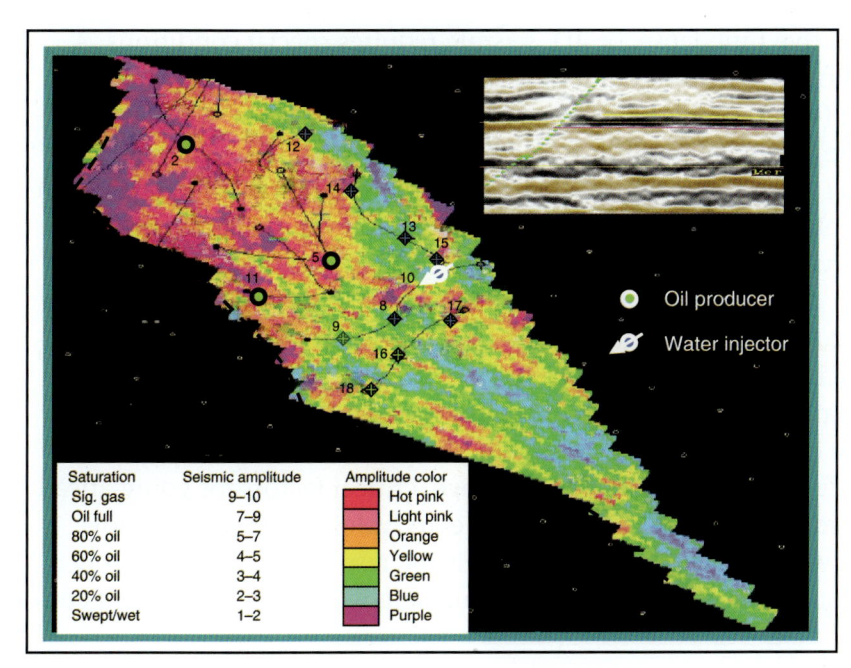

**Figure 3.** In this seismic horizon slice of a reservoir undergoing production, amplitude is correlated strongly with oil and gas saturation.

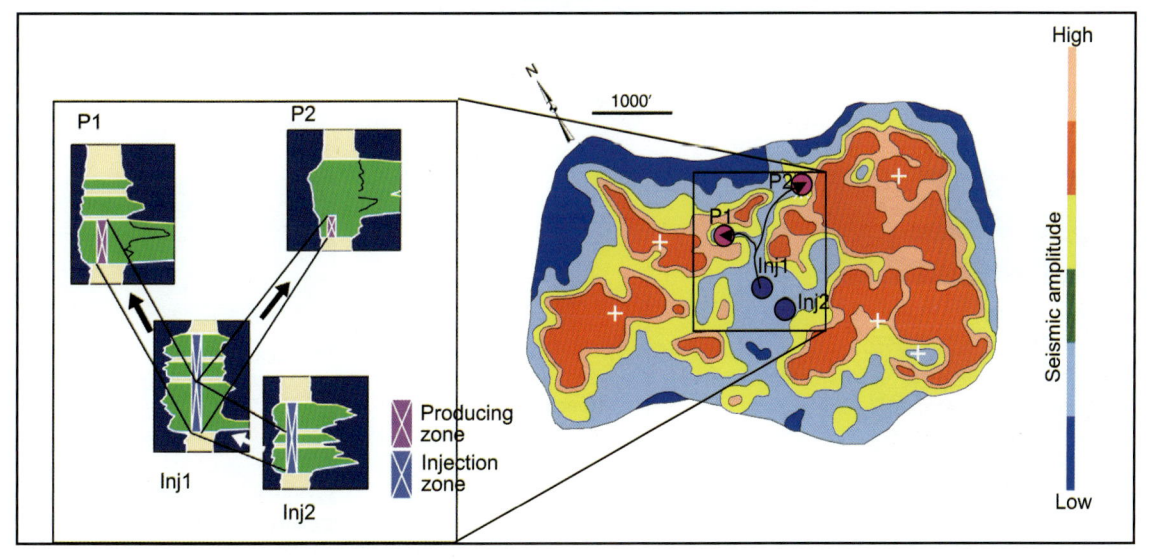

**Figure 4.** If the first 3D survey is acquired after production starts, the effects of production must be understood. This horizon slice is from a seismic survey shot 10 years into production of the fault block. Water from injection wells has reached producers but has bypassed a significant portion of the reservoir.

Simple reservoir models are easiest and fastest to generate and evaluate, but they can be very misleading. Subsurface geologic complexity reduces recoverable reserves when this complexity is not recognized or managed in the production plan. The role of seismic data in describing complexity can be very important.

### Time-lapse (4D) seismic monitoring

When multiple seismic data acquisition is conducted at different times to monitor production, a dynamic set of "pictures" of the reservoir can be obtained for use in management of the field. Effective time-lapse, or 4D, seismic requires special conditions. A workflow for implementing 4D seismic is in four distinct stages: (1) scoping, (2) project value, (3) repeat 3D, and (4) interpretation (Figure 5).

In the first stage, the necessary inputs are assembled and evaluated. The dynamic fluid-flow reservoir model is evaluated to appreciate what questions about production are being asked and what geologic impact is being considered. Often, the highest perceived value of 4D seismic involves a search for bypassed reserves. The base-case 3D seismic survey also is reviewed in detail to clarify the seismic attributes to be used in monitoring (e.g., horizon amplitude). Petrophysical models then are constructed to predict the likelihood that changes in reservoir conditions resulting from production (e.g., increases in water saturation) can be measured with repeat seismic data.

When the first stage of 4D planning is complete, the project moves to stage two, which is prediction of value of information. In that stage, critical factors for going ahead with the project are considered. They include the monetary value of 4D success (how much the bypassed oil is worth), the likelihood that seismic response can be seen if perfect seismic data were acquired, and the risk that seismic data will be imperfect (will con-

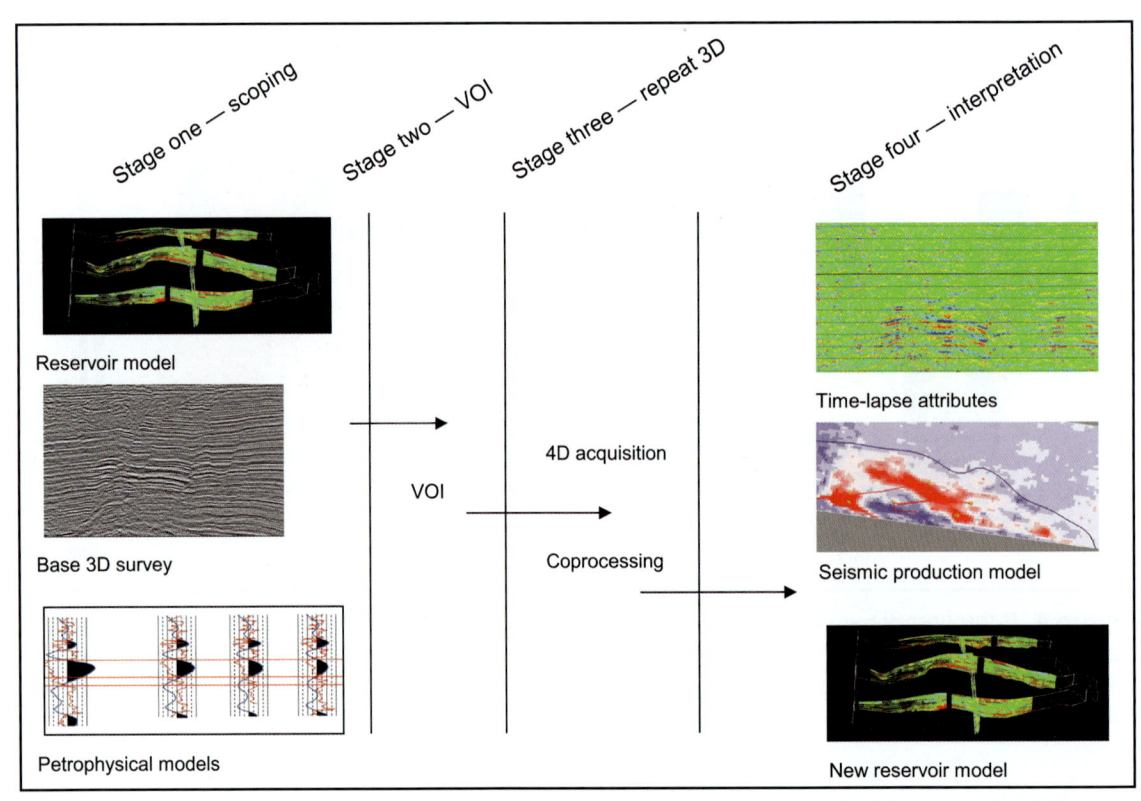

Figure 5. In this workflow for seismic time-lapse (4D) reservoir monitoring, stage one includes two principal activities: (1) assessment of the base 3D seismic survey and the fluid-flow conditions as seen from the reservoir model and (2) performing forward geophysical models to see whether production effects are resolvable seismically. Stage two is a VOI calculation made to ensure that investment is worthwhile. Stage three includes acquisition of new data and coprocessing of it along with the base survey. Stage four includes finding the appropriate seismic attributes for production effects, confirming those attributes with forward seismic modeling, and updating the reservoir flow model. The value of 4D seismic lies in supplying information from which action can be taken to improve production efficiency.

tain noise). Those factors form the basis for evaluating the investment potential of the 4D survey in a value-of-information (VOI) procedure (Figure 6). As described in Chapter 1, the decision branches are split into one that employs seismic data and one that does not. When calculated for VOI, the value of the 4D project must be greater than the cost of the seismic-data investment — otherwise, the economic model predicts that you will lose money.

When the decision is to proceed, the project enters stage three, acquisition and processing. It is preferable to acquire a base 3D survey prior to production, and the base must be of sufficient quality to measure reservoir properties. If possible, an attempt should be made to reacquire the 3D seismic survey with acquisition parameters identical to those of the base survey. That stipulation is especially important for offset ranges, source-receiver azimuths, and total fold of the data.

After acquisition, the critical step of seismic data processing proceeds. It is important to process the new acquisition along with the base-survey data so that time, amplitude, and phase are matched as exactly as possible outside the affected reservoir zone. Normally, that procedure requires not only the processing of the new acquisition but also a simultaneous reprocessing of the base survey. The objective is to produce seismic data with the same response in areas not affected by production and different response in areas affected by production.

With stage three completed, stage four (interpretation and evaluation) begins (Figure 5). Time-lapse seismic data are reviewed for the expected change in attributes predicted in stage one. For successful 4D, differences in the attribute(s) observed should have resulted only from production. Seismic time and amplitude changes are the most widely measured attributes. They are expected to be zero outside the affected production zone.

However, expectations should be managed by an understanding of actual seismic observations. Not only can amplitude of reservoir reflectors be changed by differing fluid saturation during production, but pressure changes can have impact as well. That is especially true when reservoirs are at or near bubble point and gas is coming out of or going into solution. Seismic reflections showing time delay or advances often are observed because of changes in fluid saturation and pressure. In addition, some variations in seismic response can be caused by reservoir compaction during fluid extraction.

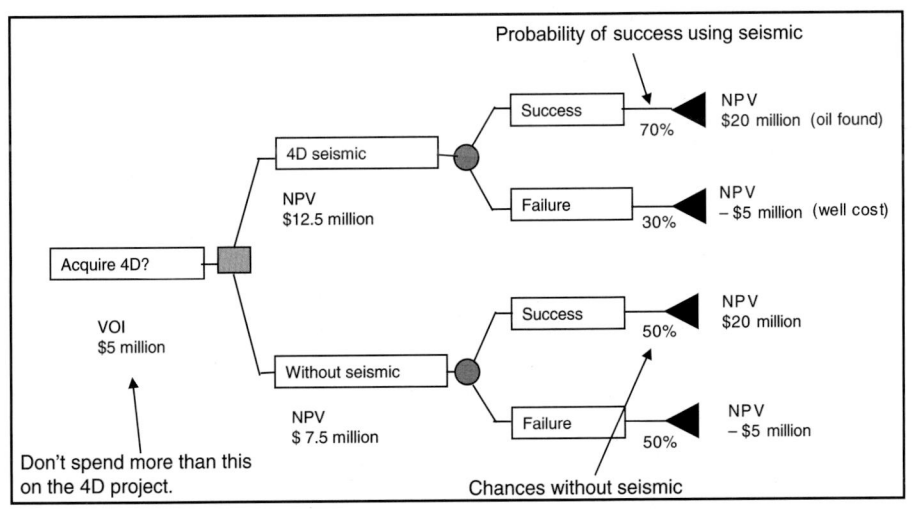

**Figure 6.** In this sample VOI calculation for a 4D project, the model says the value is $5 million. That amount is high but might not be enough to support the most expensive acquisition and processing procedures available.

Many of the seismic effects observed in 4D projects are more complicated than intuition might suggest. Therefore, it is important to perform forward geophysical models to explain observations and resolve ambiguity. The way to do that is to employ the reservoir-engineering fluid-simulation models at the time step of the seismic surveys. Using the appropriate geophysical parameters, forward models of seismic data are produced with the attempt to match seismic observations. In that way, the reservoir-engineering model can be updated from seismic observations. By understanding the reservoir dynamics and "seeing" the production effect between wells, the reservoir can be managed for best value.

Four-dimensional seismic data have been applied adequately only in the past decade or so. Therefore, surprises still occur during projects. You might try to monitor bypassed oil and instead find that you are seeing effects of an expanding gas cap or even reservoir compaction. Some level of value for surprise (unanticipated reservoir changes) should be built into the question of investing in 4D seismic for production management of a reservoir with no prior 4D acquired.

Despite the still recent use of 4D applications, they have been used in many production environments and have been valuable in many cases. In oil-producing fields, 4D data have provided new information for interpretation of water influx and channeling that has left bypassed reserves. Such data also have provided valuable information about the compaction of the reservoir and the expansion of the encasing rock. That is a direct observation of the reservoir drive mechanism and has led to recognition of unproduced compartments.

In heavy-oil environments, 4D data have shown the impact of steam flooding on the reservoir, including the heating of oil, released gas, and direction of steam movement. Cases also are reported in which $CO_2$ injection into carbonate reservoirs has been monitored with repeat seismic surveys. Therefore, under the right conditions, 4D seismic data can provide information that is useful for understanding reservoir dynamics.

In most geologic environments where seismic signal is good, 4D seismic should be considered for production monitoring. If seismic traveltimes and/or amplitudes respond to fluid changes, the reservoir is a good candidate for evaluation by 4D seismic.

## 4D seismic example — Girasol, Angola

Dubucq et al. (2003) present a textbook case for time-lapse seismic data. In the West African Gulf of Guinea, the Girasol field was an oil discovery in stacked deepwater turbidite channel sands. Two delineation wells were drilled before the additional 11 development wells were drilled and put into production. Reservoir modeling was used to evaluate the risk of uneven reservoir drainage caused by stratigraphic variations and flow barriers. Production of the reservoir requires water injection in some wells and an additional gas injector at the top of the reservoir for pressure support.

High-quality 3D seismic data were acquired for field development. Three years later, after production had begun, the same acquisition parameters were used to acquire the time-lapse survey. The two surveys were processed in parallel to produce the difference, which shows the effects of oil production, water injection, and gas injection (Figure 7).

Interpretation of the 4D seismic-data results shows several important dynamics of reservoir production. The horizon slice of the 3D base data set is a preferred interpretation attribute because the amplitude shows stratigraphic variations, reservoir properties, and potential flow barriers (Figure 8). For the 4D difference data, the same attribute section shows effects of stratigraphic barriers, gas-injection connectivity, and unswept oil.

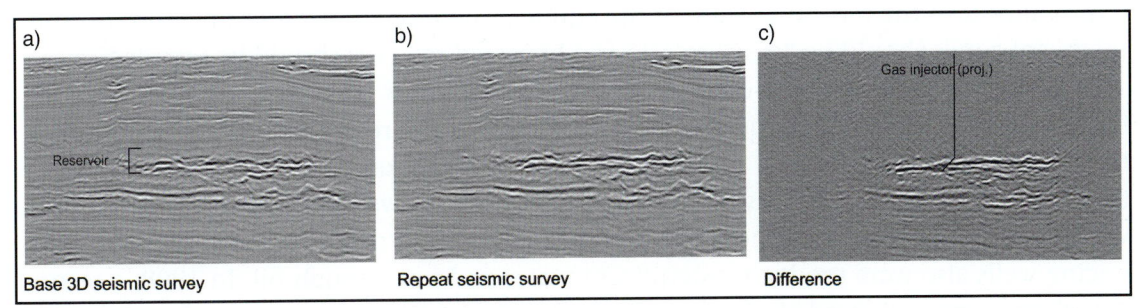

**Figure 7.** Shown here are (a) a base seismic 3D survey, (b) repeat survey for monitoring, and (c) the difference between the two. The gas injector introduces an amplitude difference in the reservoir. Because the gas also lowers the velocity, the time delay in the repeat survey causes the differences in the time position of the reflectors below the reservoir (Dubucq et al., 2003).

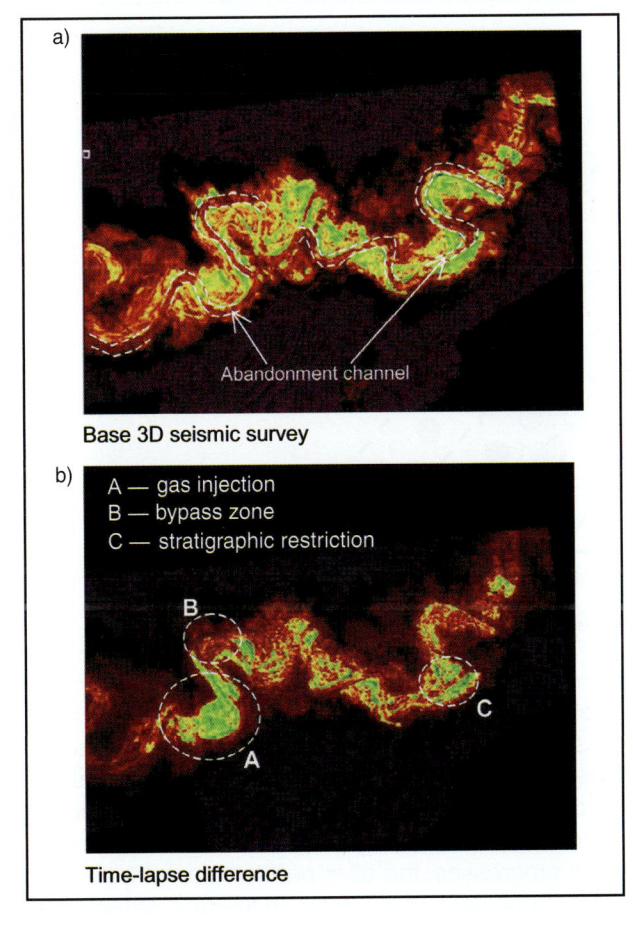

**Figure 8.** Shown here are (a) the horizon slice (amplitude) of the base 3D survey and (b) the 4D difference. In Figure 8a, note the high-amplitude meanders of the channel system (green represents highest amplitude). An interpreted abandonment channel of lesser amplitude cuts through the system. In Figure 8b, the difference map, the highest amplitudes are associated with injected gas. Site A illustrates the high-amplitude location of the gas injection, B shows an area of oil bypassed by injected gas, and C shows the effect of the abandonment channel (Dubucq et al., 2003).

## 4D seismic example — Bay Marchand, Gulf of Mexico

Let's study some additional examples from the Bay Marchand field of the Gulf of Mexico. Figure 9 shows data from wells that define the petrophysical response for a water-wet sandstone reservoir at 7100 ft; this reservoir has a measurable reflection where it is encased in the surrounding shales. The oil sands have significantly higher reflection, and this generates an oil bright spot. However, when oil is produced, the depleted zones return almost to the reflection state of the aquifer. Therefore, during seismic monitoring, the swept zones should be "easy" to see, and bypassed reserves should be revealed.

This project had no base-case 3D survey. In 1987, after 30 years of production, the first 3D seismic survey was acquired. A horizon slice of the reservoir shows high amplitude where producing wells A, B, and C are making oil (Figure 10). In 1987, an additional observation well (D) was oil full. Repeat logging shows that by 1999, the reservoir was swept significantly, and the water level in the well had risen. By 1999, the three producing wells also were producing significant water and not enough oil. In 1998, a second 3D seismic survey showed a seismic response consistent with water encroachment in the wells. The high-amplitude area showing unproduced oil indicates that producers are drawing up the waterfront in this relatively low-viscosity oil.

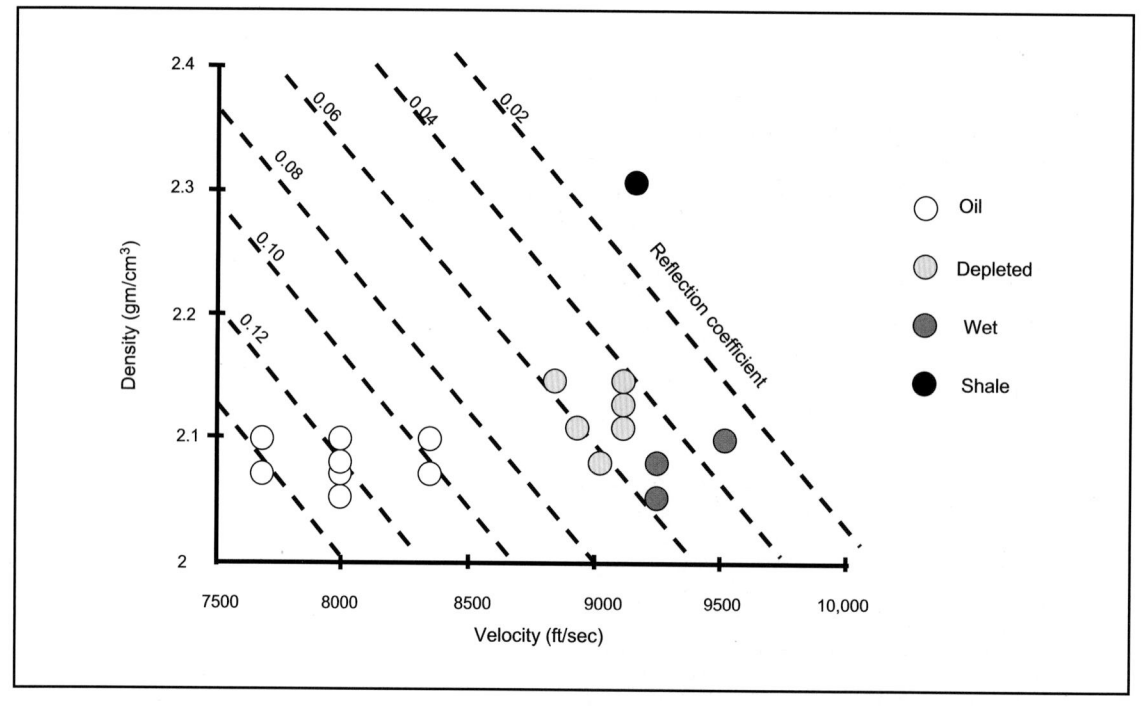

Figure 9. This reservoir example is a Tertiary shelf sandstone in the Gulf of Mexico. Acoustic properties from well logs show a low reflection response (amplitude) where the reservoir is water wet and reflection coefficients are 0.03–0.06. Where the reservoir is oil saturated, the high gas content of the oil causes a high-amplitude seismic response. This generates reflection coefficients from 0.10 to 0.14. That is two to three times larger than the water-wet response and constitutes an oil bright spot. When production of the reservoir progresses, the oil is partly replaced with water, and the reflection coefficient diminishes to low numbers comparable to the fully water-wet case. Repeat 3D surveys should permit the monitoring of production effects of that magnitude easily.

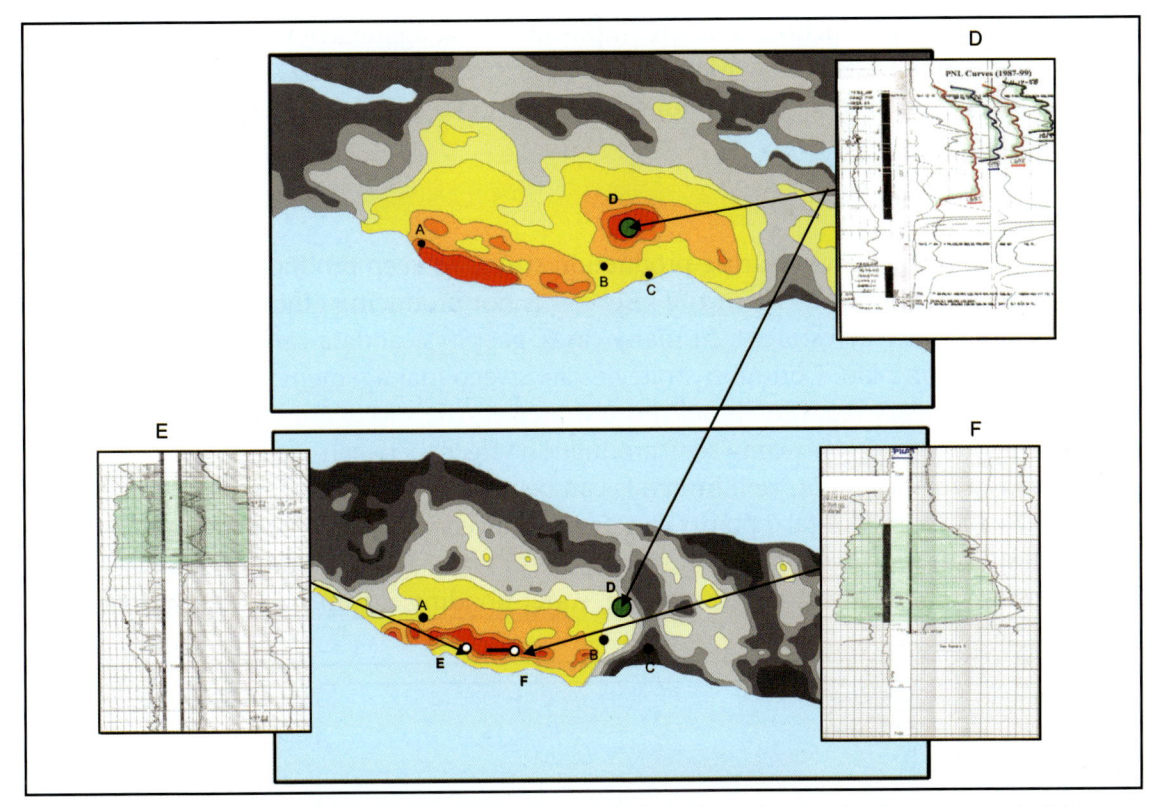

**Figure 10.** Bay Marchand field, Gulf of Mexico. Horizon slices of two seismic surveys shot 11 years apart show oil movement from bottom-water drive in the reservoir. The three producing wells (A, B, and C) began to make significant water by the time the second survey was shot, as can be inferred by the lowering of amplitude adjacent to the wells. Influx of water and lowering of seismic amplitude are confirmed by observation well D. Bypassed reserves were indicated in the remaining high-amplitude areas. Those reserves were confirmed during the drilling of directional well E and produced by horizontal well F.

However, the new survey also showed a high-amplitude area directly between wells, suggesting uneven reservoir sweep and bypassed reserves. Follow-up drilling by Chevron proved that to be a correct interpretation and led to the completion of an additional horizontal development well which added good production rates to the field.

In a separate fault block, a similar but more complex result was observed at the 7600-ft level. The repeat survey showed the oil sweep was very uneven because of low-viscosity oil and the vertical and horizontal interconnectivity of reservoir facies (Figure 11). From 1987 through 1998, wells G, H, and I produced significant volumes of oil. Wells G and H were making significant water cuts by 1998, and well I had watered out completely. Again, the second seismic survey revealed bypassed oil reserves. Wells were drilled by Chevron in cooperation with Energy Partners, Ltd., in all three high-amplitude zones of the 1998 survey and proved bypassed reserves. Two were put onstream, adding to production.

The prior examples of 4D impact represent just part of the impact seismic data have had on the entire field. The impact of seismic data on the overall management of the Bay Marchand field was significant. The first 3D survey identified unrecognized fault blocks, stratigraphic complexity, and bypassed reserves. Improved management of the field and increased production were accomplished by an aggressive program of infill drilling, horizontal wells, and workovers (Figure 12). The second 3D survey revealed evidence of additional production dynamics and additional opportunities for further sweep management.

## Remarks

Reservoir management during production strives to keep production up while keeping costs down. That means when the reservoir is not producing efficiently, intervention might be a necessary investment. In many cases, geophysical data can be of assistance in the intervention plan. Common strategies are sweep management, pressure maintenance, and secondary recovery. In all those strategies, the successful use of 4D seismic data can help the reservoir team "see" production effects between wells.

In sweep management, well intervals can be choked back if they are drawing too much water or additional take points can be introduced with more perforations or further infill drilling. For pressure maintenance, gas or water injectors can be introduced. For secondary recovery, reservoir fracturing, water flooding, steam injection, or $CO_2$

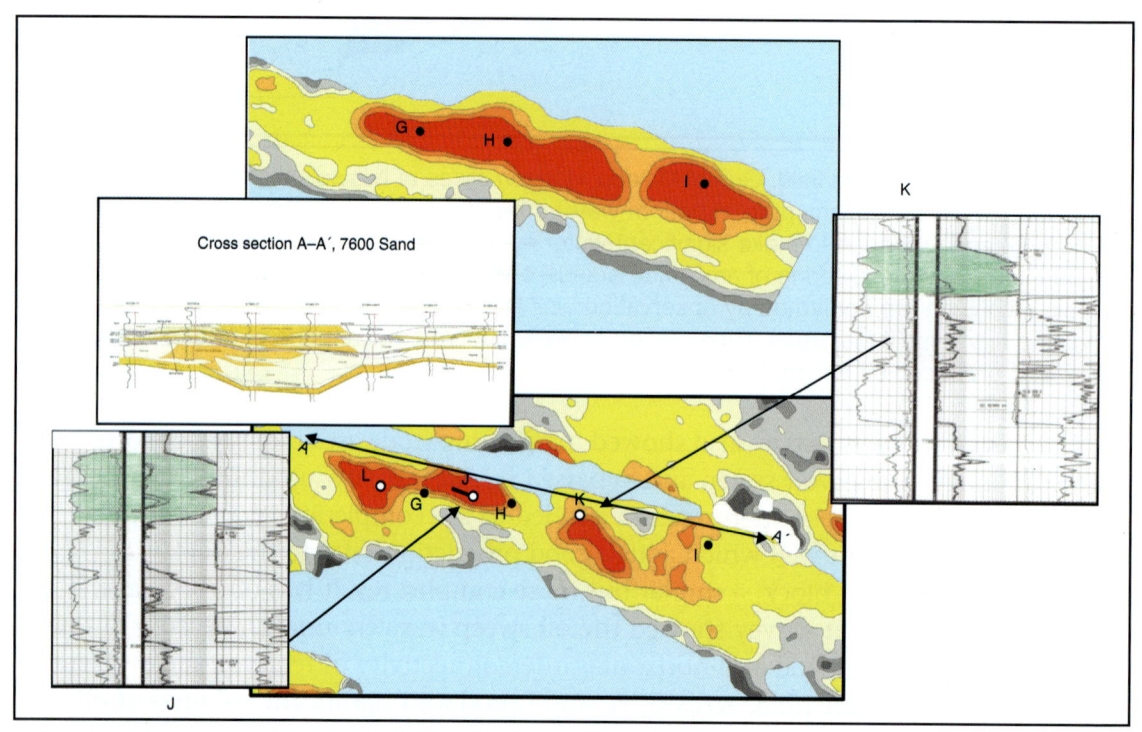

Figure 11. Bay Marchand field, Gulf of Mexico. Sandstone facies of the deltaic reservoir are interconnected, leading to complex bottom-water drive. In an adjacent and deeper fault block than that shown in Figure 10, 4D monitoring also provided evidence of bypassed reserves. Wells J, K, and L were drilled to define and produce the bypassed oil.

injection can be introduced. Although intervention procedures add costs, the benefits can be very rewarding. Geophysical data add value to that part of the oil-field project by providing direct measurements of production effects among wells.

Although seismic reflection data are the dominant tool employed in reservoir geophysics, additional important geophysical technologies add value in reservoir management. Although not discussed here in detail, examples show direct value is added from application of crosswell seismic, crosswell electromagnetics, gravity measurements, and passive seismic recording.

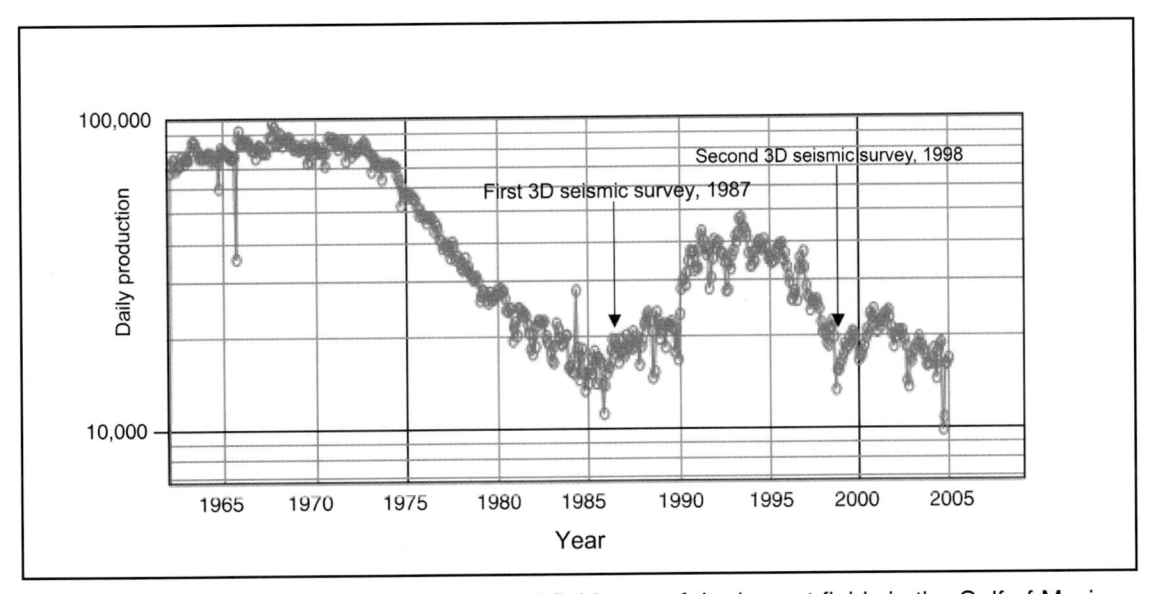

Figure 12. Production rates of Bay Marchand field, one of the largest fields in the Gulf of Mexico. After the drilling of more than 500 wells, production from the field went into steep decline. However, production was revitalized with infill drilling, improvements in water-flooding efficiency, and selective horizontal-well production. The improved field management was a direct consequence of applying the interpretation from a full-field 3D seismic survey directed at production and development.

# Chapter 6  Development Problem

### Objective

The problem described here is intended to illustrate how development and production proceed over long periods of a field's history. It is intended as a tool for teaching the impact of seismic data. As with other real examples, some things about this case are somewhat inconsistent, incomplete, or imperfect. However, it is worth the time to follow the problem and become involved. Here is what you can expect:

1) Given key information concerning a production case history that requires reservoir management, define where to drill three or four development wells.
2) Be prepared to discuss the development-program design and the reasoning behind it.
3) There is no "right answer" to the problem, but the actual results of the development problem will be shown.

### Key information

You have just joined the offshore team and have been placed on the Cobra project. Lucky you! The production department is planning to accelerate production in one of the best producing fault blocks in the field. The field has been producing mainly oil for 18 years and is in secondary recovery through water flood. With oil prices having increased lately, accelerating production should make a good profit.

The plan is to do some infill drilling to put about three more producers in the remaining oil zones to maximize recovery. *The task has fallen to you to position the new producers in the fault block.* Although the concept seems simple enough, the production response of the fault block has not always worked as predicted, and there are complications to deal with. You will want to know the project background, so you find the files and talk to the current team of engineers, geologists, and geophysicists.

### Background geology

First you find out where the project is located. In that area, the Tertiary section produces prolifically from shelf sands. The shales are the source and seal, and the reservoirs are faulted sandstones. The structure map (Figure 1) shows that the fault block you are assigned is about 4 km × 2 km and apparently isolated by faults in all directions. Seismic data have been used to map fault blocks and to site wells. However, because the seismic data are 2D recordings, their quality and density are limited, and the data have been used primarily for structural information (Figure 2).

**Figure 1.** The reservoir of interest is bounded by faults. Water is only about 100 ft deep, and the shoreline is to the northeast.

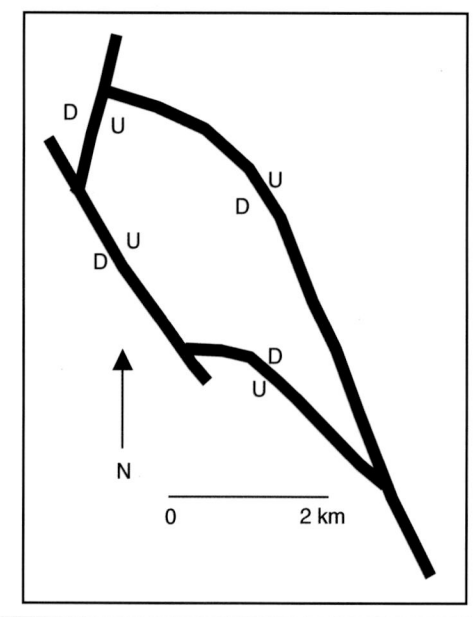

**Figure 2.** A seismic line that covers the fault block of interest shows low dip structure between 1 and 2 s.

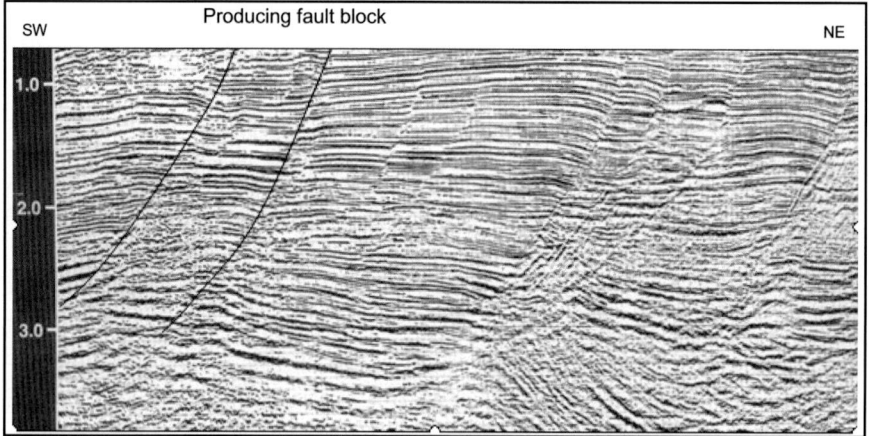

An important element of field management has been to understand the geologic complexity of the reservoir. The delta in which the reservoir was deposited was very rich in sand. Reservoirs are stacked vertically and extend over large horizontal distances. Therefore, well logs can be correlated, and sequence stratigraphy can be interpreted (Figure 3). Because the deposition was deltaic, reservoir properties vary horizontally, primarily according to the original depositional facies. Predominant geologic facies recognized from well and seismic data are delta-front sands, barriers, bars, channels, and overbank deposits.

Some reservoirs produce a great deal more than others. Concentrate on the E-1 reservoir unit, the largest producer in the fault block (Figure 4). To understand how best to locate development wells today, see how the reservoir has behaved over time.

## Discovery, delineation, and development

You are briefed by the team on the discovery and development history of the fault block and become better acquainted with how it has been developed. This helps to provide clues to where your next producers should be drilled to obtain more oil production.

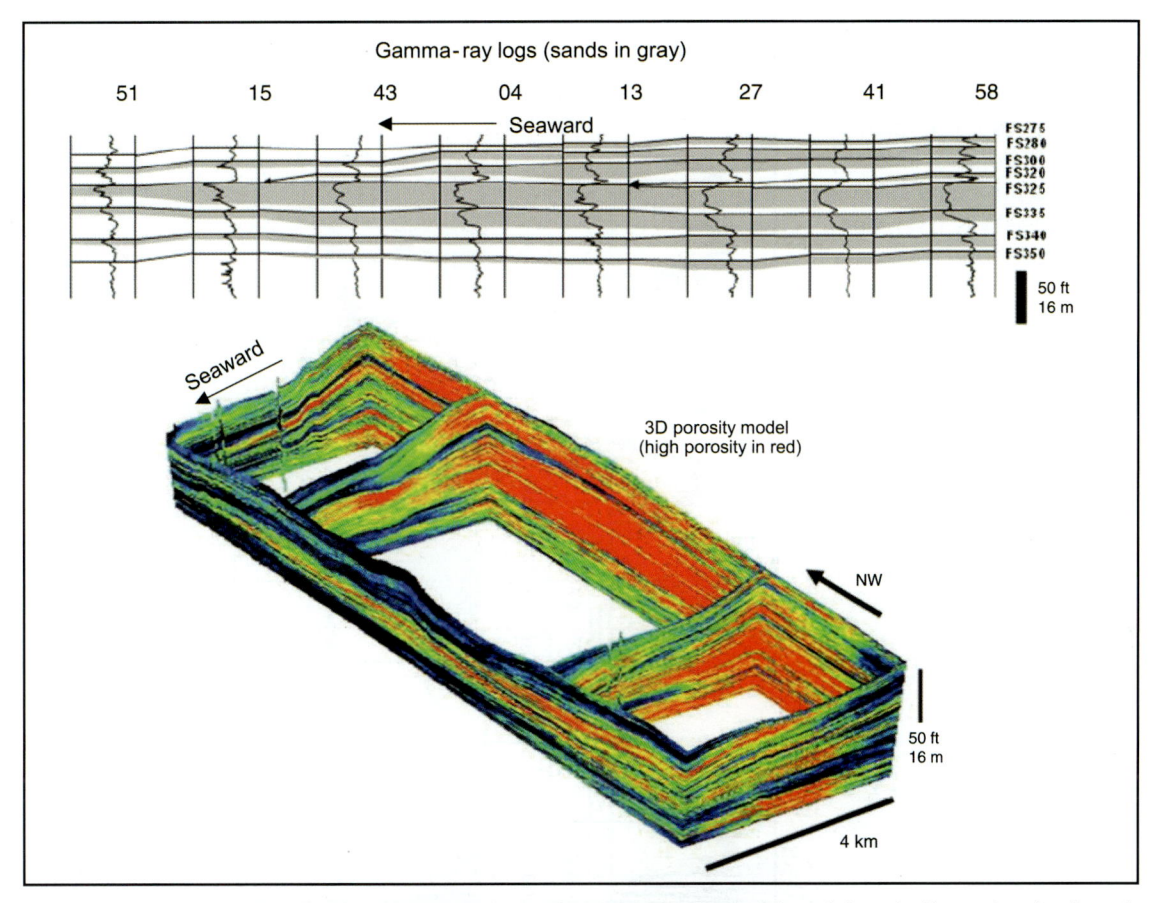

**Figure 3.** A cross section of gamma-ray logs shows the connectivity of the shelf sand units (gray). This includes facies from tidal to delta front. Sand units are separated by flooding surfaces and are overlain by sealing shales. The fence diagram of the units, showing high porosity in red, illustrates high connectivity of sands shoreward and reduction of porosity seaward.

Well #1 discovered gas in the E-1 sand of this fault block on the top of the structure (Figure 5). In the following year, delineation drilling discovered that gas was underlain by oil in wells #2 and #3. In the well farthest downdip to the northeast (#4), the reservoir was oil full, suggesting that the oil leg also could extend downdip to the southeast (Table 1).

In that same year, another delineation program targeted the gas-oil contact but found wells #5 and #7 to be oil full. Updip well #6 established the location of gas-oil in the southwest (Figure 6). With that additional information, the reservoir had been delineated, and development could start. In the second year, the full potential of the reservoir was established with development wells #8, #9, and #10 (Table 2), which were drilled even farther downdip to the southeast. They established the position of the oil-water contact. In the third year, well #11 was drilled as a development well to balance production. The well was successful in drilling at the gas-oil contact; only 9 ft of gas column was penetrated and was completed in the oil section below. Six wells were put into production in the third year (Figure 7). The reservoir produced significant oil in the primary phase of production from pressure depletion and edge-water drive (Table 3).

**Figure 4.** A typical log from the field shows reservoir sand in yellow, oil in green, and gas in red. The main log curves defining lithology and fluids are shale volume (VSH), resistivity (RT), density (RHOB), and neutron (NEU). The E-1 reservoir has a gross thickness of at least 100 ft and a net-to-gross sand ratio of more than 80%.

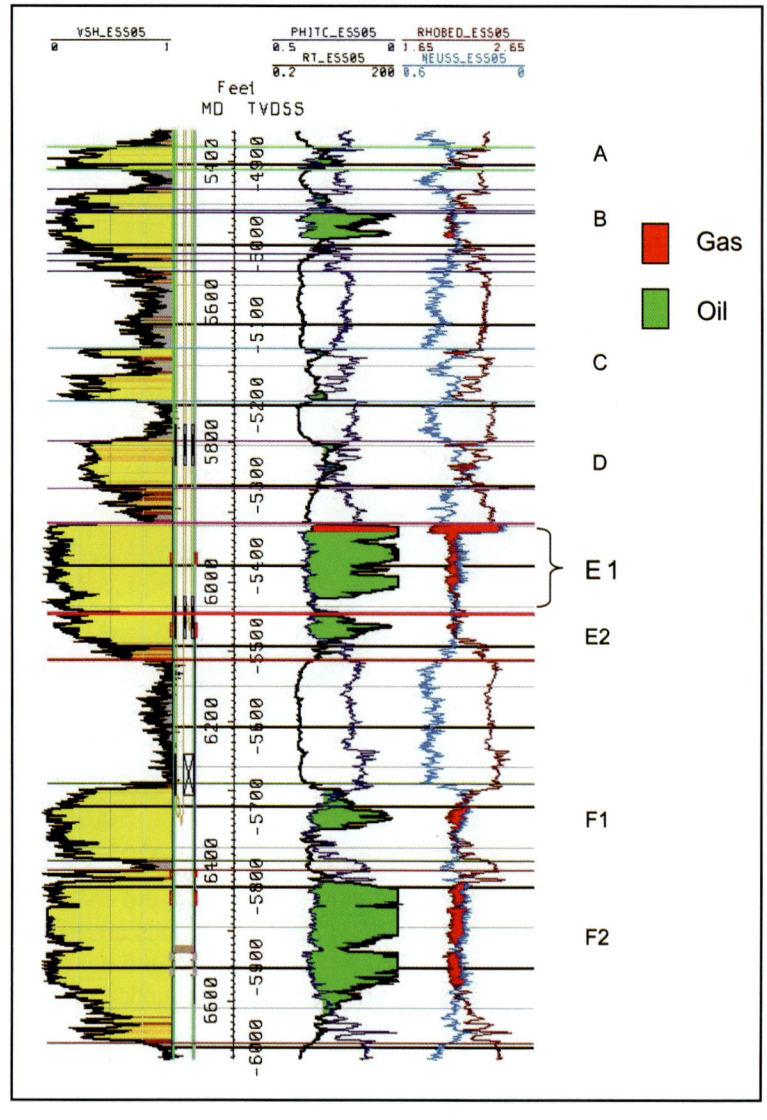

**Figure 5.** The fluid map shows oil (green) and gas (red) from the discovery phase of the reservoir. Depth contours are 100 ft. Discovery well #1 was drilled at the top of the structure. Delineation wells discovered oil farther downdip.

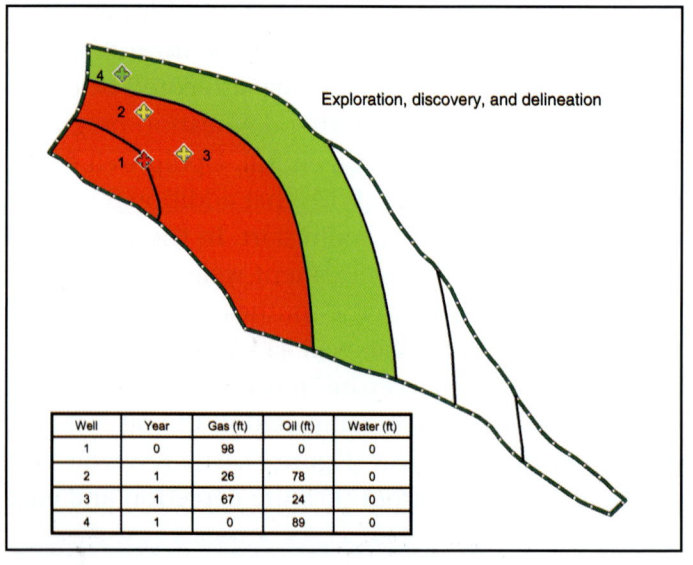

| Well | Year | Gas (ft) | Oil (ft) | Water (ft) |
|---|---|---|---|---|
| 1 | 0 | 98 | 0 | 0 |
| 2 | 1 | 26 | 78 | 0 |
| 3 | 1 | 67 | 24 | 0 |
| 4 | 1 | 0 | 89 | 0 |

Table 1. Discovery and early delineation wells of the E-1 reservoir during year one.

| Well | Date of drilling | Gas (ft) | Oil (ft) | Water (ft) |
|---|---|---|---|---|
| 1 | Year one | 98 | 0 | 0 |
| 2 | Year one | 26 | 78 | 0 |
| 3 | Year one | 67 | 24 | 0 |
| 4 | Year one | 0 | 89 | 0 |

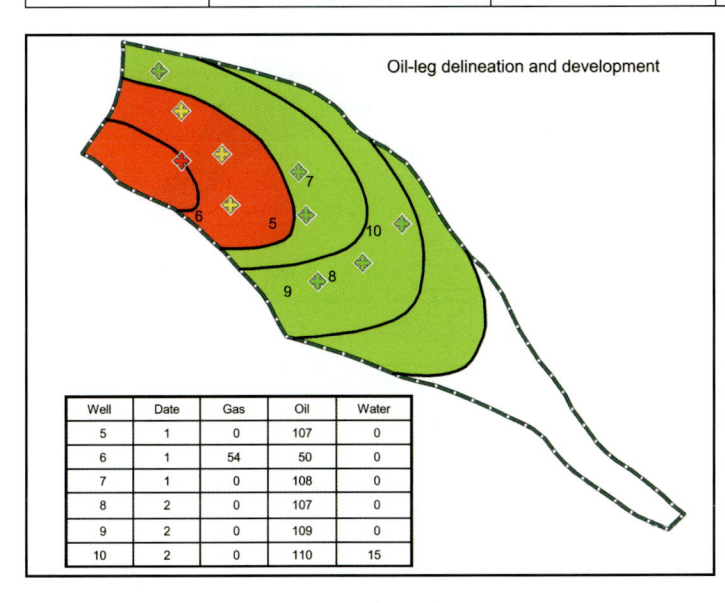

Figure 6. The fluid map from the second delineation phase shows the modified interpretation of structure and the extension of discovered oil to the southeast. Most development wells drilled downdip did not encounter any water and were oil full. Well #10 established the position of oil-water contact.

Table 2. Delineation and development wells drilled in E-1 during years one and two.

| Well | Date | Gas (ft) | Oil (ft) | Water (ft) |
|---|---|---|---|---|
| 5 | Year one | 0 | 107 | 0 |
| 6 | Year one | 54 | 50 | 0 |
| 7 | Year one | 0 | 108 | 0 |
| 8 | Year two | 0 | 107 | 0 |
| 9 | Year two | 0 | 109 | 0 |
| 10 | Year two | 0 | 110 | 15 |

## Production

From year three to year five, oil was produced from the reservoir, and the oil-water contact moved updip. To accelerate production, three additional production wells (#12, #13, and #14 ) were put onstream (Table 4; Figure 8). Production continued for the next five years until water broke through in the northeastern part of the field (Figure 9). Wells #10, #13, and #14 produced so much water they were taken offstream. Primary production continued until the fourteenth year, when reservoir pressures had decreased to the point that additional support was required.

In its initial production, the E-1 reservoir showed that it behaved like many deltaic clastic reservoirs that are managed for high initial flow rates. A time line of the activity (Figure 10) shows a rapid delineation and development strategy, with production startup within three years and further production infill within an additional two years. Primary production was mostly uninterrupted for eight years before production declined substantially.

**Figure 7.** When the first six wells were put into production, the reservoir produced primarily oil, with advancement of water from the southeast.

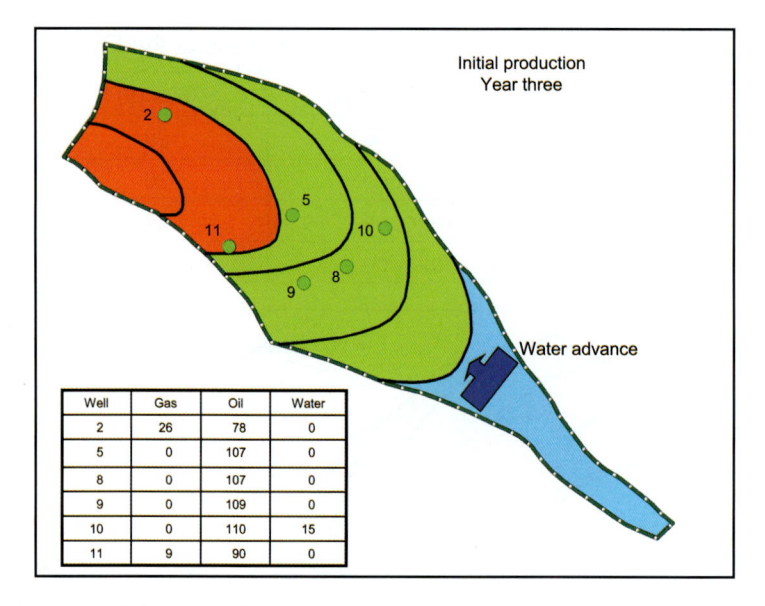

Table 3. Production wells in the E-1 reservoir by year three.

| Well | Date | Gas (ft) | Oil (ft) | Water (ft) |
|---|---|---|---|---|
| 2 | Year one | 26 | 78 | 0 |
| 5 | Year one | 0 | 107 | 0 |
| 8 | Year one | 0 | 107 | 0 |
| 9 | Year one | 0 | 109 | 0 |
| 10 | Year two | 0 | 110 | 15 |
| 11 | Year three | 9 | 90 | 0 |

Table 4. Development wells drilled in the E-1 reservoir in year five.

| Well | Date | Gas (ft) | Oil (ft) | Water (ft) |
|---|---|---|---|---|
| 12 | Year five | 0 | 122 | 0 |
| 13 | Year five | 0 | 110 | 0 |
| 14 | Year five | 0 | 103 | 0 |

**Figure 8.** The map of fluid from year five shows that infill wells were drilled updip and away from the advancing waterfront.

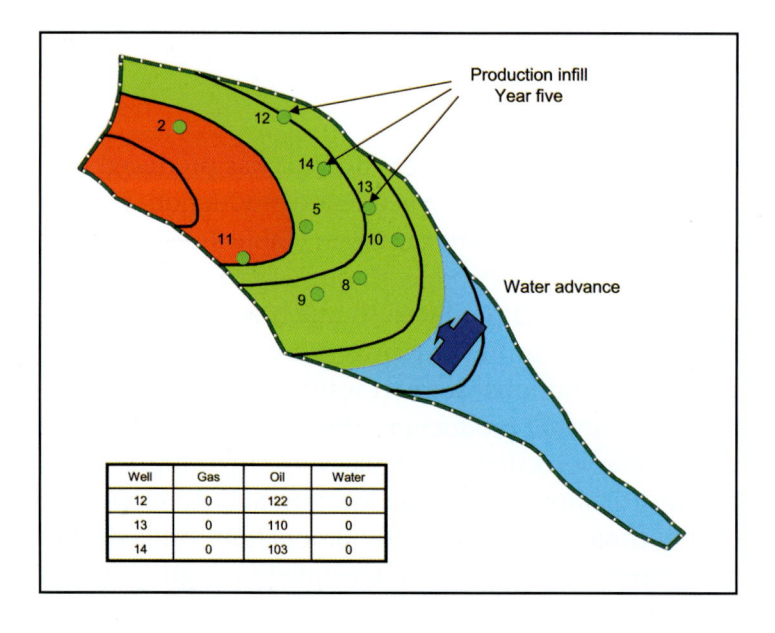

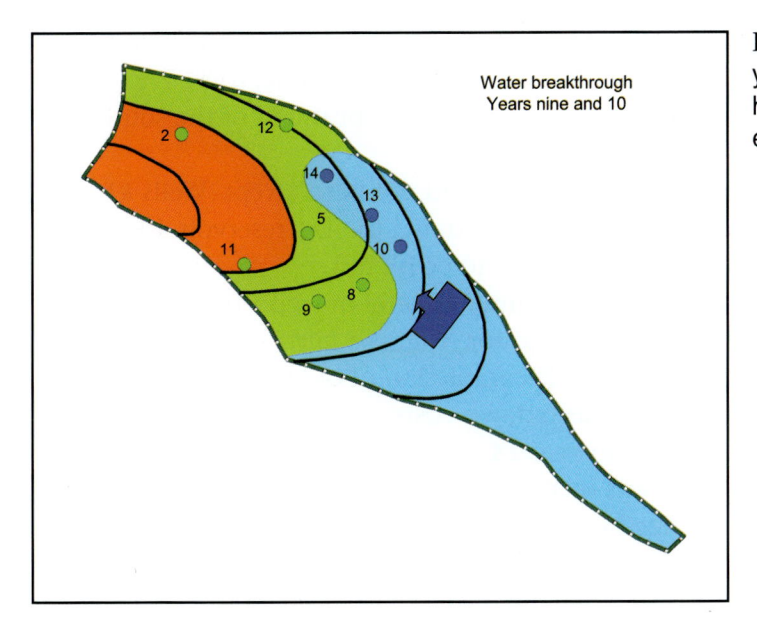

**Figure 9.** The map of fluid from year 10 shows that the waterfront has broken through in the northeast.

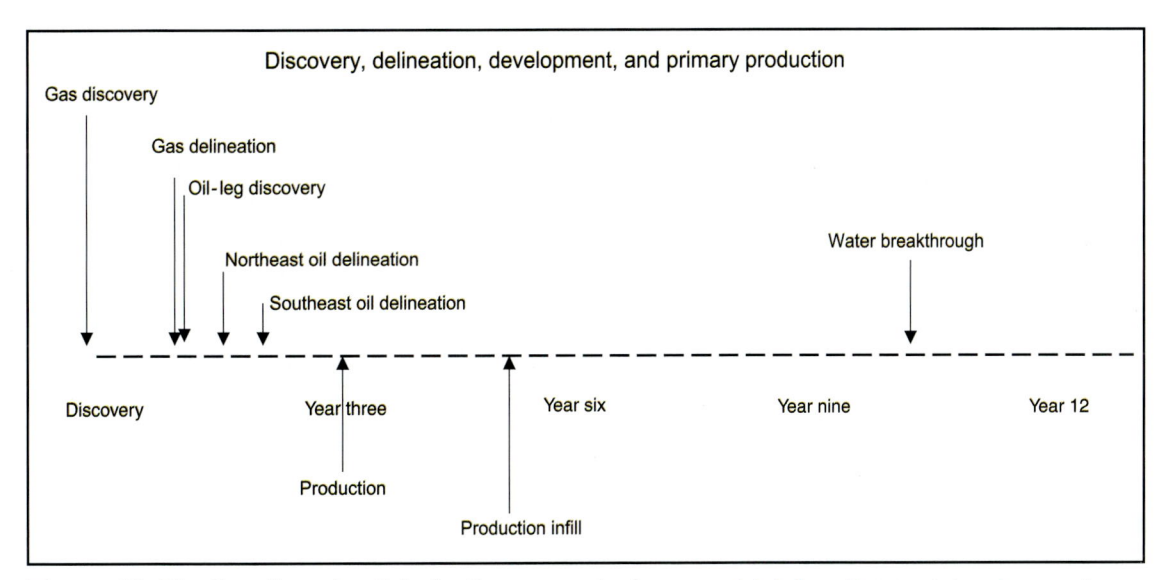

**Figure 10.** The time line of activity for the reservoir shows rapid delineation and development. The reservoir was brought into production in only a few years. Within eight years, water had begun to break through to the producers.

## Secondary recovery

Because of the drop in reservoir pressure and the opportunity to have a more efficient sweep, in the fourteenth year, water injectors were planned downdip, in the water leg. By injection of water, a more even sweep of the reservoir can be obtained, and bypassed oil can be moved. In addition, water injection increases reservoir pressure over an even larger area, which results in a larger production volume from the reservoir.

Water injectors placed at the edge of the oil-water contact provide maximum pressure support and sweep. Two injectors were considered at the same depth interval because well #10 had watered out (Figure 11). The presence of water moving into the northeast was confirmed by well #15, which was drilled for deeper reservoirs. Well #17, also drilled for deeper reservoirs, confirmed that water had swept much of the southeastern area.

However, in the southwestern area, the reservoir had not been swept. Wells #16 and #18, which were drilled for deeper reservoirs, penetrated the E-1 reservoir. Instead of being swept by water, the wells showed the reservoir to be oil full.

For secondary recovery, water injection was initiated only in well #10, adjacent to well #15, turning an older producer into an injector. That is a fairly common way to manage facilities for optimizing production. Recompletion of old wells for injection makes use of existing facilities, which helps keep down costs.

When well #10 began to inject water in year 14, six updip producers were kept onstream. Oil production from the east was increased significantly. By year 18, the water from injector #10 had reached to the northeast, and well #12 had watered out. Other wells (#8 and #9) also had gone offstream, so only three wells were left in production (Figure 12).

Producing from only those three wells might not be the optimum recovery program for the reservoir. That is why the team is given the task of recommending locations of additional production wells.

## Impact of secondary recovery

The history of production rate in this fault block is similar to the production of the entire field. It is illustrated by the annual graph of produced oil and water (Figure 13). Primary production increased as development wells were put onstream. Notice that after initial production, the field went into strong decline until an important turnaround occurred. Secondary recovery was initiated with water injection, and production increased to nearly the initial volume. That improvement was obtained at a price, however. Water production also had increased from movement of the aquifer, increased interstitial production, and recovery of injected water.

Figure 11. The map of fluid from year 14 illustrates the uneven advance of the waterfront as wells in the northeastern part of the field began to produce more water. The southwest had not been swept as well, as confirmed by additional penetrations. Secondary-recovery methods were instituted by water injection in well #10.

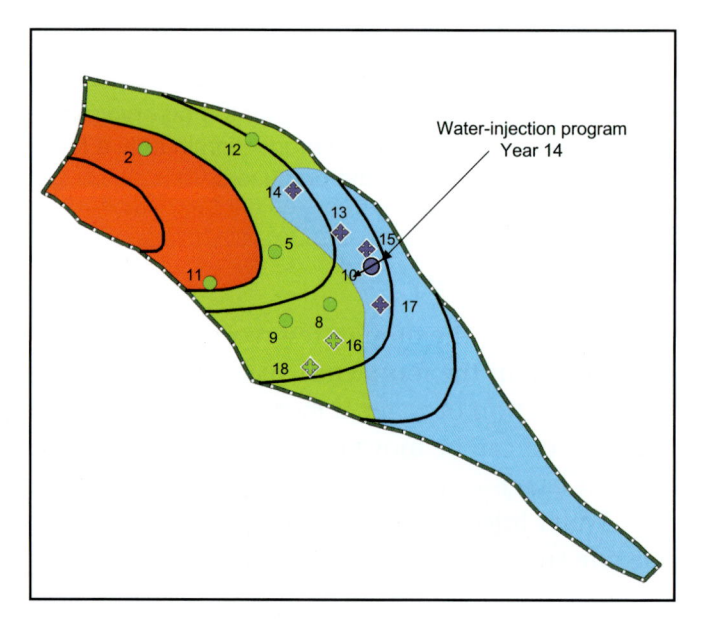

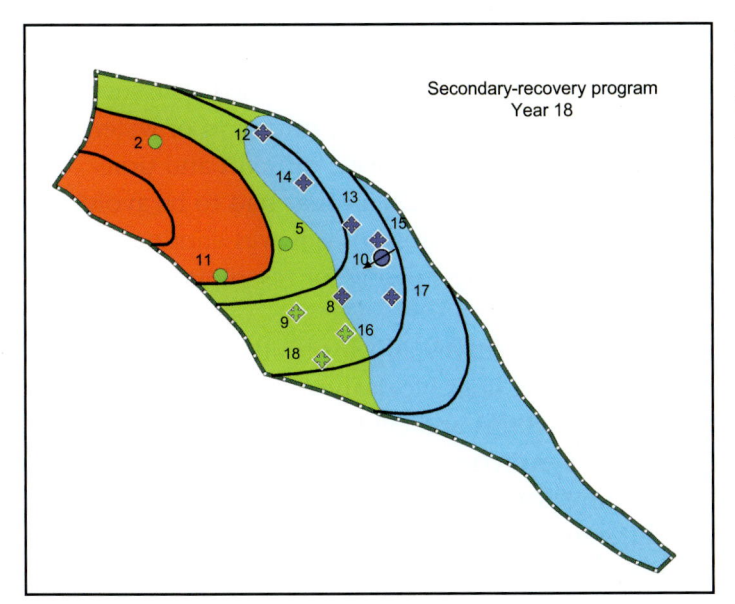

**Figure 12.** The map of fluid from year 18 shows water had broken through in the northeast into well #12.

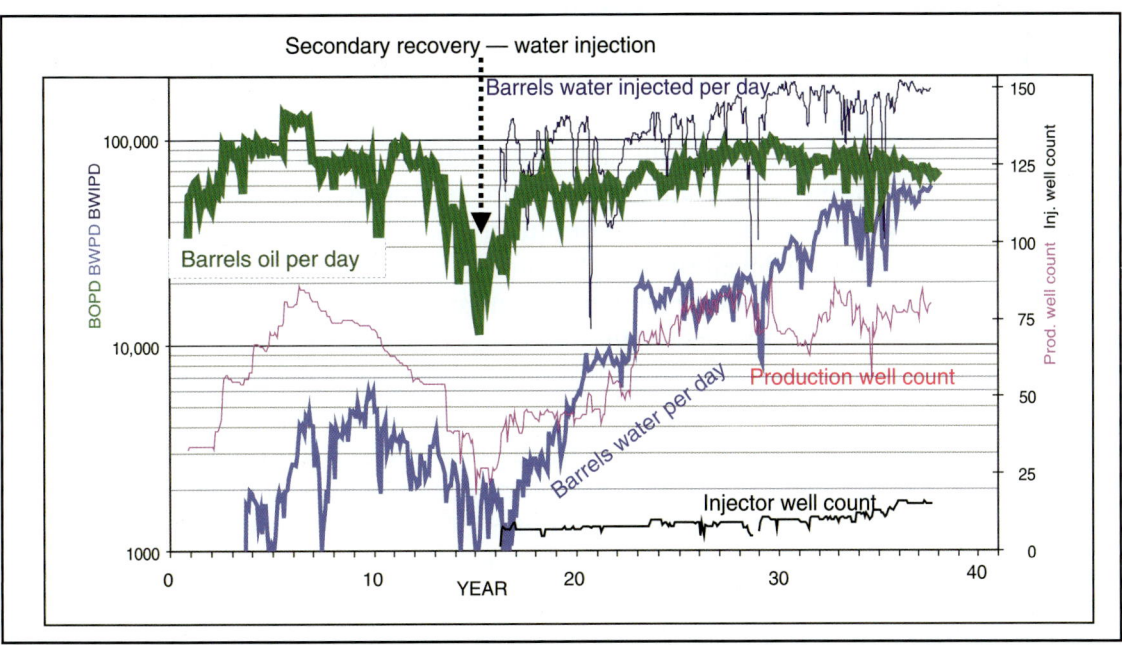

**Figure 13.** The field production behaves much like the eastern reservoir alone. After initial high oil production (green), decline in pressure causes steep production decline. Then water injection and infill drilling revive production. However, total water produced (blue) increases rapidly.

Now you have reviewed the history of the reservoir and can discern something about the dynamic state of its production. Although water seems to be advancing generally updip, the sweep across the reservoir is not even. Water apparently has moved in comparatively large volumes along the reservoir's northern-bounding fault. Areas to the south are not swept as much; they might contain some bypassed oil. Based on the reservoir as you see it at this point, where would you consider drilling additional wells?

## Properties of the 3D seismic survey

The first 3D seismic survey over the field recently has been processed. The survey was acquired 18 years after initial production and four years after initiation of water injection (Figure 14). Acquisition and processing of the data were difficult and time-consuming because of the large number of obstacles in the field that had to be avoided. Noise from producing facilities was recorded during acquisition; this was eliminated by special processing. Despite those problems, the data are suitable for reservoir geophysical analysis.

A review of the seismic data suggests that the survey appears to illuminate the reservoir successfully. A seismic cross section over the fault block shows stacked reservoirs. You can see both the direct evidence of hydrocarbons and evidence of the potential for reservoir-property measurements (Figure 15). Synthetic seismic traces generated from well logs show that acoustic impedance from reservoir sand is not generated from the velocity response but is dominated by differences in densities of sand and shale (Figure 16). In fact, the correlation is so strong that the density log appears to be a very close match to the gamma-ray log. Thus, the acoustic response from seismic data is an excellent proxy for lithology.

An inverted product from seismic can be used to predict the presence and quality of sand between and beyond wells (Figure 17). Very low impedances are expected from high-porosity reservoir sands; they will have better reflectivity standout. Thus, when seismic data are inverted, the seismic-attribute data are very likely to provide the evidence for prediction of sand presence and reservoir quality.

Seismic data show the location of high-porosity sand, which could indicate reservoirs. Furthermore, discrimination of the type and saturation of fluids might be possible also. Petrophysical modeling has been done to help understand whether the seismic amplitudes are good predictors of fluids. This work (Figure 18) shows that when the reservoir sand is saturated with gas, oil, or water, it has measurably different responses. This reservoir has a gas-oil ratio (GOR) of 400. Where the reservoir is gas full (80% gas) compared with oil full (80% oil), the seismic amplitude decreases 30% or more. As water fully sweeps the gas-saturated oil to a residual saturation of 20%, the amplitude decreases yet another 40%. So if the quality of the seismic data is good enough, amplitude attributes can be used to locate areas of high-porosity reservoir and possibly to discriminate the present fluid content.

## Interpretation of the horizon slice

The geophysics group has mapped the reservoir horizon of the E-1 in the fault block and has extracted the amplitude for your inspection. Compare the interpreted oil map showing the potential presence of fluids (Figure 19a) with the seismic horizon slice (Figure 19b). Observe the striking fact that the updip end of the reservoir is of much higher amplitude. That correlates well with the map of remaining oil and gas. That observation is encouraging, but some highly anomalous parts of the horizon slice do not fit.

Figure 19 is an interpretation; it might be oversimplified. The seismic-amplitude map shows considerably more spatial variation. Recall that the seismic-data amplitudes contain some noise. Also recall that porosity affects seismic response as well as fluid

*Development Problem*

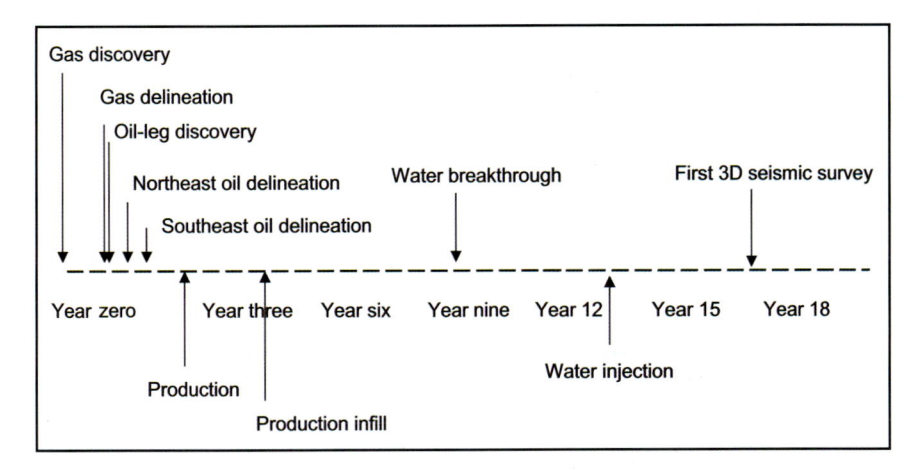

**Figure 14.** A time line of the project shows the date of the first 3D seismic survey, initiated 15 years after initial production.

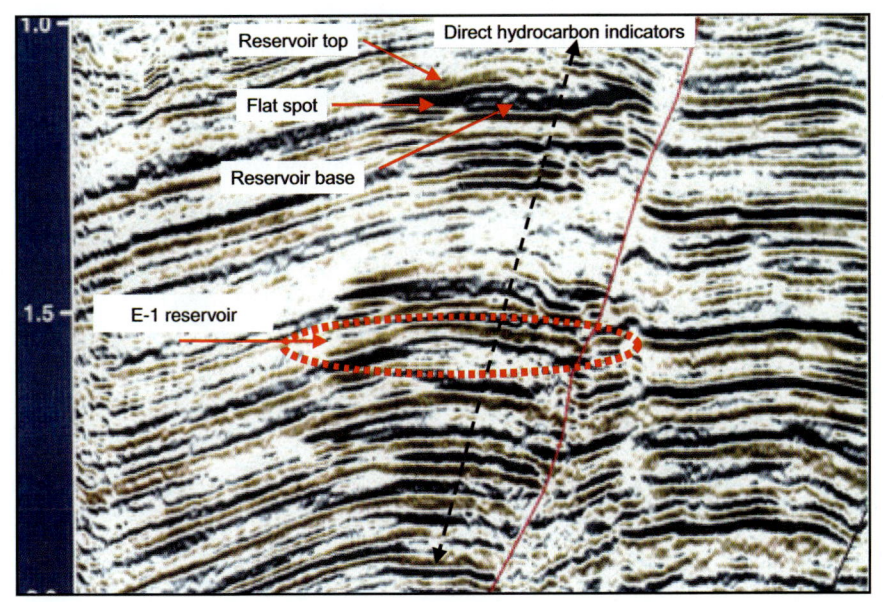

**Figure 15.** The seismic cross section of the fault block of interest shows high-amplitude direct-hydrocarbon anomalies stacked from 1.1 to 1.9 s.

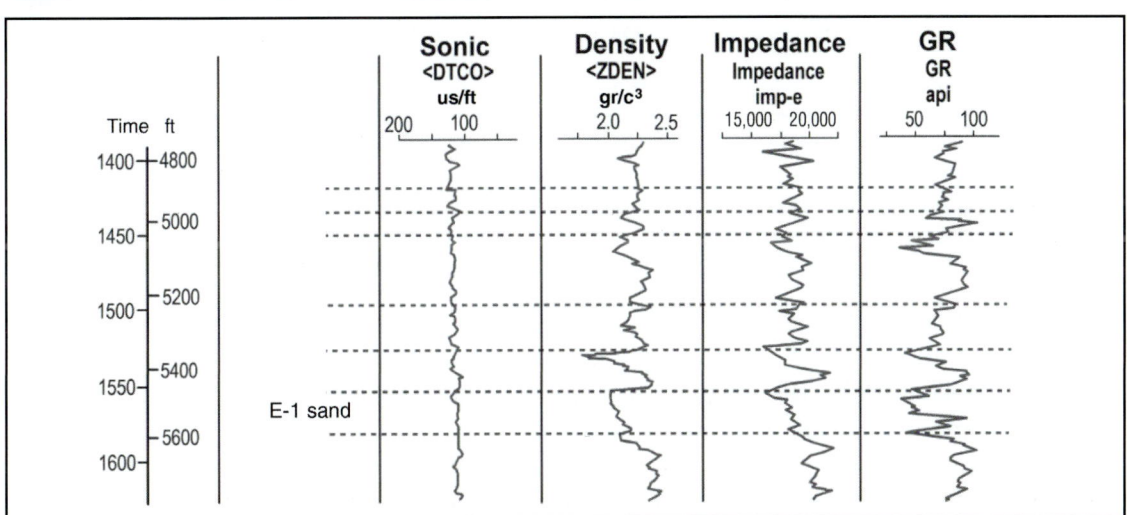

**Figure 16.** Logs of a typical well in the development blocks show high correlation among density, impedance, and lithology. Velocity (sonic) has almost no effect on impedance, which provides the seismic response.

Reservoir Geophysics: Applications

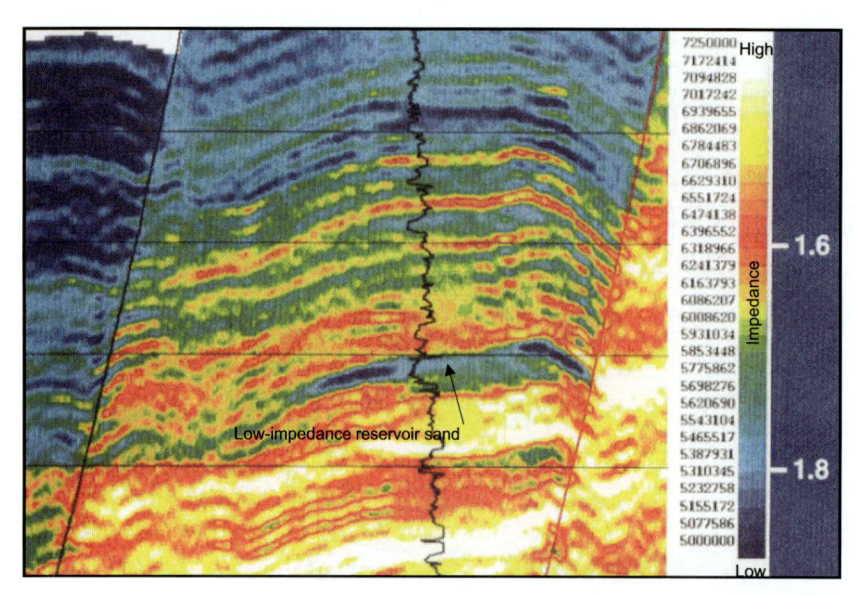

Figure 17. A relative-impedance cross section generated from the high frequencies of the seismic data and the low frequencies of the log data shows how geology can be interpreted from inverted data. Note how the low-impedance sand (blue) is thickest at the right on the downside of the fault. The sand thins toward the left away from shore.

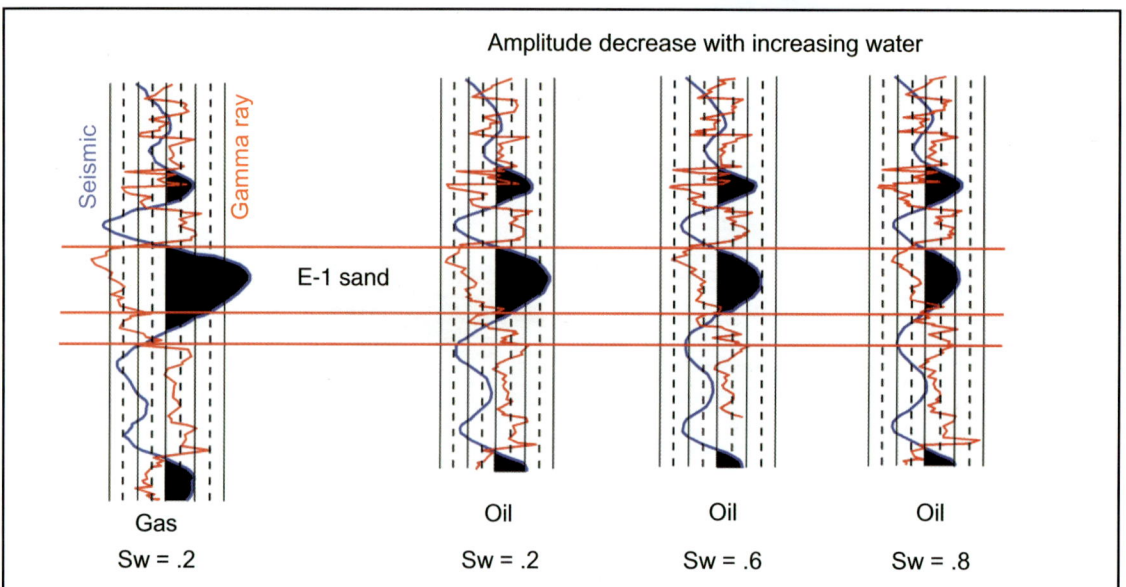

Figure 18. The seismic reflection from the reservoir is sensitive to fluids. Gas gives the highest amplitude (height of the black peak). Oil produces a lesser amplitude and water the least. Amplitude variations are approximately linear, so differences in oil saturation (1-Sw) can be mapped from horizon slices. (Note for advanced geophysicists: The synthetic traces shown are phase-shifted by −90° and are polarity-reversed. In simple reflection systems, this is a trick that mimics inversion.)

type and saturation. Therefore, although we wish that the amplitude map would show only the effects of fluids, the seismic response is affected by at least four factors, which include uncertainty caused by noise.

Nevertheless, the trend looks promising, and you plot the location of producers and injectors, calibrating seismic amplitude to fluid and saturation (Figure 20). If the amplitudes are good indicators of oil and gas saturation, then the seismic data confirm that water has not swept the reservoir in an even front. The interpreted oil map (Figure 19) shows evidence of water moving updip along the northeastern boundary fault; the seis-

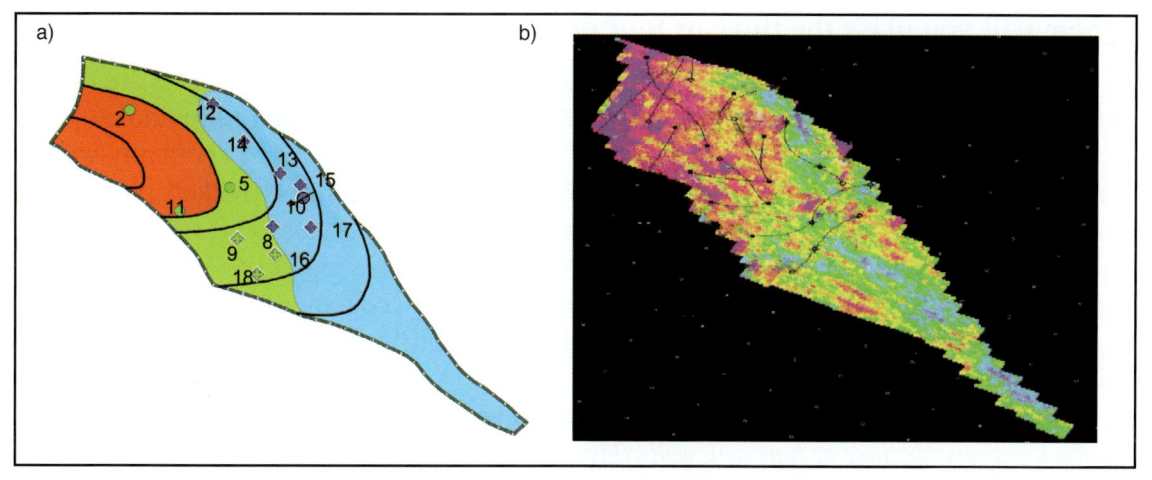

**Figure 19.** (a) The fluid map of year 18 is compared with (b) the amplitude map of the 3D seismic survey acquired at that time. The suggestion is strong that high-amplitude areas might not have been swept with water and might hold hydrocarbons.

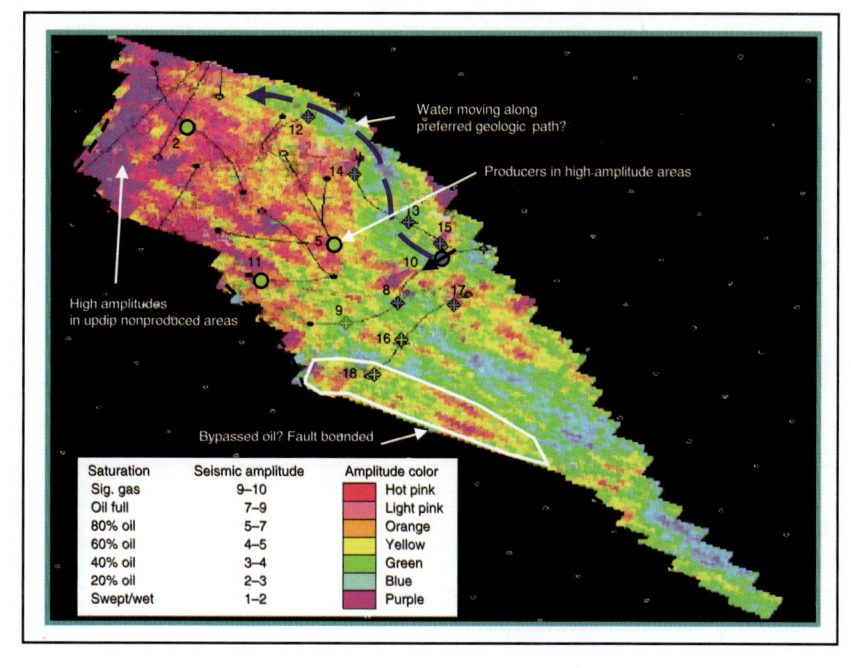

**Figure 20.** The seismic horizon slice (amplitude map) of the E-1 reservoir shows locations of the three producing wells in high-amplitude areas (red). Calibration of fluid saturations to seismic amplitude is a fairly good fit. It shows connected low-amplitude areas where updip wells have watered out. The suggestion is strong that the waterfront is channeling to the northeast. Downdip, some areas still have patchy high amplitudes, which might be evidence of bypassed oil.

mic data (Figure 20) also show that evidence. The three remaining producers making oil with high saturation are all in high-amplitude areas. That is consistent with the oil map in Figure 19a as well.

However, the seismic response is complicated in or near the water zone. Either the seismic amplitudes are not correlated well to fluids in that area or local stratigraphy is complicated or both. Geologic heterogeneity would have resulted in uneven reservoir sweep and some pay might have been bypassed, as suggested by patches of high amplitude where water should be dominant. The case for some bypassed oil is supported in part by the high amplitudes in a connected "fault block," containing well #18, which drilled through the E-1 reservoir to deeper targets. Well #18 is downdip of the injector and below the nominal water contact, but the oil column was 66 ft thick. The oil might have been bypassed partly because of a small fault within the reservoir.

## Where will you place the three or four new development wells?

Before you spot the new development wells on the amplitude map and write down the $x$ and $y$ coordinates, remember the purpose of the wells. They are to provide oil production for as long as possible in the highest quantity. You will want to avoid any water influx from water channels. You also will want to stay away from the thickest part of the gas cap, or the wells will produce gas, but not enough oil. Wells that produce both gas and oil are tricky to complete and to operate correctly because mobility of gas is greater than that of oil. Summary economics of the development program are favorable for drilling infill wells only if the wells produce significant amounts of oil and small amounts of water for several years (Table 5).

*Now look at the amplitude map (Figure 21) and write down the coordinates of your choice for the locations for three or four new wells (Figure 22).* Justify the location of each well in terms of the expected oil, gas, and water. How long will the wells produce before they must be shut in for making too much water?

## Results

Once you have decided where to drill development wells, you can find documentation for the remainder of this case in Appendix B. There you can follow development and production of the reservoir during its next 10-year period.

Table 5. Investment opportunity for development.

| What you need to invest | |
|---|---|
| Development well costs (4 × $10 million) | $ 40 million |
| G & A | $ 2 million |
| **Total additional project costs** | **$ 42 million** |

| What you might earn | |
|---|---|
| Resource added per well | 4 million Bbls |
| Revenue $20 Bbl × 4 wells | $ 320 million |
| Less well costs | $ (42) million |
| Less operating expenses | $ (40) million |
| Before-tax income | $ 238 million |
| **After-tax (50%) income** | **$ 119 million** |

Development Problem

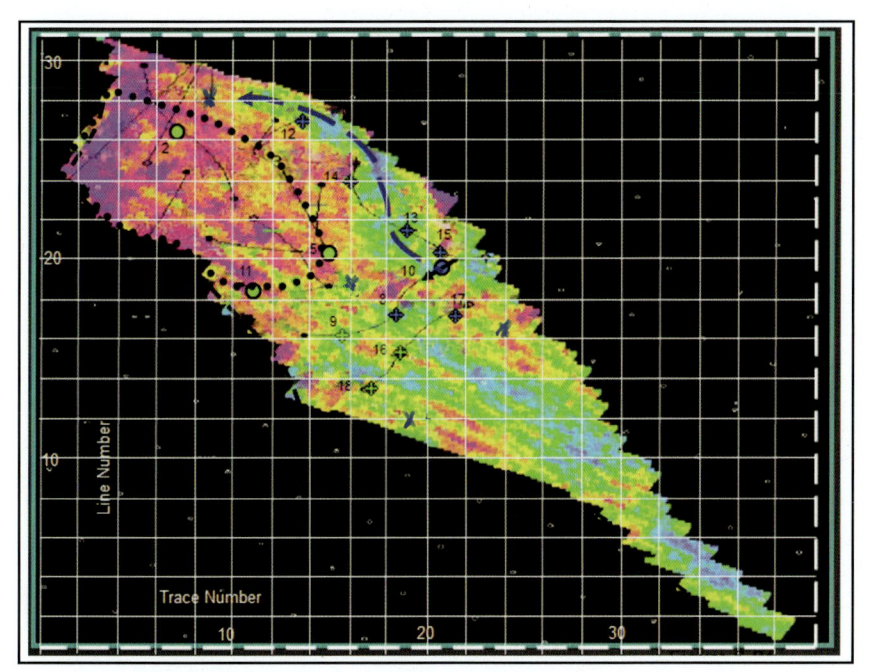

**Figure 21.** This horizon slice (amplitude map) is for helping you choose locations for development wells. An estimate of the location of the present day gas cap is represented by a black dashed line. Use the line and trace index on the axes to record the coordinates of where you want to drill wells intended to accelerate production.

Choose locations for three or four additional wells for development purposes (write down line and trace location numbers).
- Maximize oil production.
- Minimize water influx.
- Stay out of the gas cap?

| Well | Line | Trace |
|------|------|-------|
| A | 12 | 19 |
| B | 19 | 17 |
| C | 28 | 9 |
| D | 17 | 24 |

**Figure 22.** Use this table for tabulating the line and trace index of your choices of well locations. Consider three or four locations and put them in order of priority.

# Chapter 7   Heavy Oil

Geophysics is challenged in some very important producing environments, most notably in fields that produce heavy oil and in fields with carbonate reservoirs. Those environments have different economic and technical requirements that tend to make the use of geophysics less common. However, because those environments are so important to the hydrocarbon production of the world, there has been consistent work to improve geophysical applications, some of which will be covered in the following sections.

## The heavy-oil business

In what way does heavy-oil production differ from light-oil production, and what role does geophysics have to play? Heavy oil has the property of high viscosity (thickness), which means it will not flow easily to production wells. In general, high viscosity is caused by biodegradation of lower-viscosity oils. Bacteria digest high-fraction hydrocarbons and leave the heavier fractions behind, so the oil becomes like tar. Large heavy-oil deposits normally are found near the surface because of the difficulty of survival for bacteria at the high temperatures encountered at depth. Some oil fields near the surface do not have good seals and will allow bacterial degradation. Light-fraction hydrocarbons also are lost because of evaporation.

So how did all the oil get close to the surface? In the world's two largest deposits, in western Canada and eastern Venezuela, the answer is that oil migrated updip out of the basin, and where it became biodegraded, it left behind a giant "tar mat" (Figure 1).

What are the implications of high viscosity on oil-production techniques? Vertical wells drain very small areas when oil viscosity is high. Thus, it is necessary to drill many wells close together or to penetrate the reservoir with multiple horizontal wells.

For example, the Kern River field in the San Joaquin valley of central California is approximately 24 square miles and contains 10,000 producing wells. On average, each well drains an area smaller than a football field. Westward on the other side of the valley, well spacing is four times as dense. In the heavy-oil fields of Venezuela, oil has slightly higher viscosity and flows more slowly. As a result, horizontal wells are drilled from a single vertical, and extensive penetration from many multilaterals is used to provide closely spaced completions.

Production techniques for heavy oil include some of those used in conventional reservoirs plus some especially adapted to slow production rates. Most heavy-oil fields are supported by secondary-recovery methods early in the life of the field. Primary production does not recover a high percentage of oil in place in heavy-oil fields, and secondary-recovery methods are crucial (Figure 2). It is only with secondary-recovery methods that subsurface heavy-oil fields have maintained a significant proportion of world oil production.

## Secondary recovery

What sorts of secondary methods are used for heavy-oil reservoirs? In a manner similar to conventional oil recovery, water can be pumped into adjacent injector wells. However, heavy-oil recovery from water flood often is marginal because the relative permeability of water to oil allows water to finger, break through, and dominate production volume.

A more effective method for heavy-oil production involves heating the oil in place to decrease viscosity; this results in faster and more complete recovery. Steam can be injected into a producing well and then the well can be put back on production ("huff and puff"). Steam can be injected into adjacent wells, providing heat to the reservoir and decreasing oil viscosity by as much as two orders of magnitude. Oil then can be recovered from horizontal wells. Alternately, oil can be allowed to drain by gravity and then can be produced by downdip wells in steam-assisted gravity drainage (SAGD) (Figure 3).

Production problems are associated with those techniques because so much trouble is employed to "force" the reservoir to produce. Horizontal pressure gradients from steam

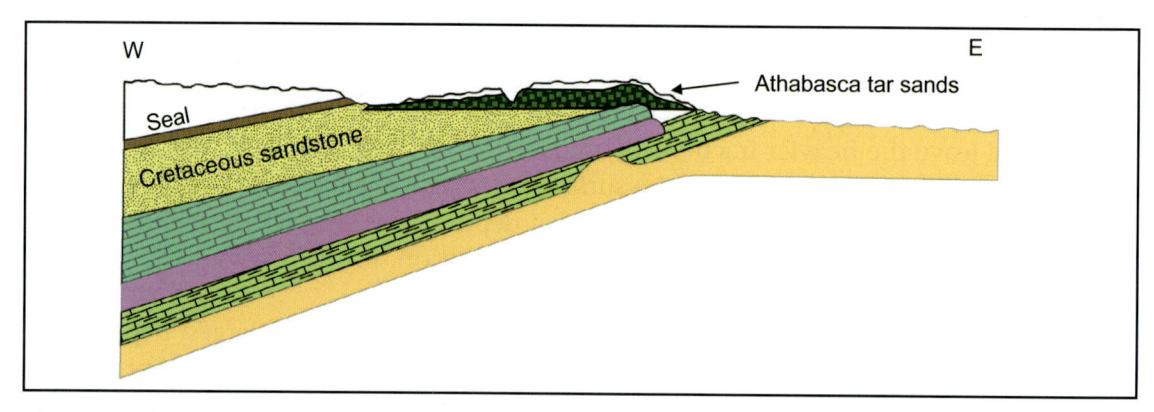

Figure 1. Heavy oil of the Athabasca originated deep in the western Canadian basin and migrated updip along Cretaceous sandstones until reaching the surface. It was biodegraded significantly, leaving behind one of the largest oil resources in the world, sometimes known as the trillion-barrel tar pit (Powley, 2007).

Figure 2. Production history of the Kern River heavy-oil field in southern California shows that in the first 50 years, primary production declined gradually to very low volume. Steam stimulation introduced in the 1960s significantly accelerated production and remains in use as the primary energy for field production.

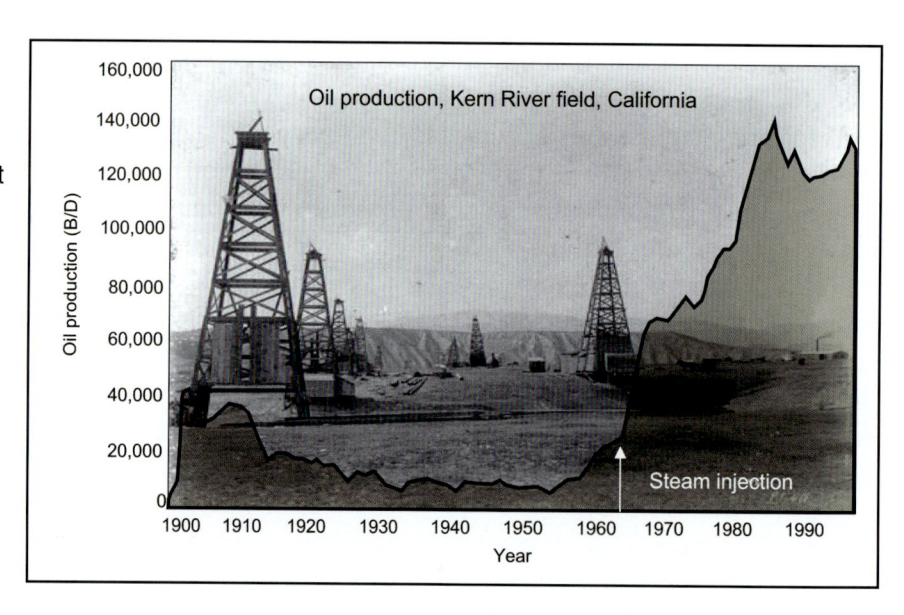

injection can be as high as 1 psi/ft. Secondary recovery in heavy-oil production also can generate thermochemical and thermomechanical problems that cause reservoir fluids to precipitate unwanted solutes, which plug production pathways or cause pipe to corrode or become brittle (Figure 4).

Mechanical problems also can occur. In some heavy-oil fields, a significant amount of sand in the reservoir is produced along with oil. Subsequent reservoir compaction can result in failure of producing facilities downhole.

In many fields, a significant amount of water often is produced during steam-injection methods. Large investments might be necessary for water handling.

## Impact of subsurface geology

The most common factor in all heavy-oil production techniques is the issue of uneven and incomplete drainage. Only a fraction of the oil can be produced with primary production and that only along connected flow units with no stratigraphic complications. During water flood and steam flood, injected waters have a strong tendency to follow shortcuts to the producing wells through high-permeability "thief zones." When breakthrough occurs, only the injection fluids will be produced in quantity because they are so much more mobile than oil. With heat-assisted production, heterogeneity in the reservoir results in considerable uncertainty about the interactions of steam, rock, and oil. Steam injection is employed in a highly dynamic environment of high heat and pressure contrasts. In addition, compartmentalization by faulting can have a substantial impact on incomplete reservoir drainage.

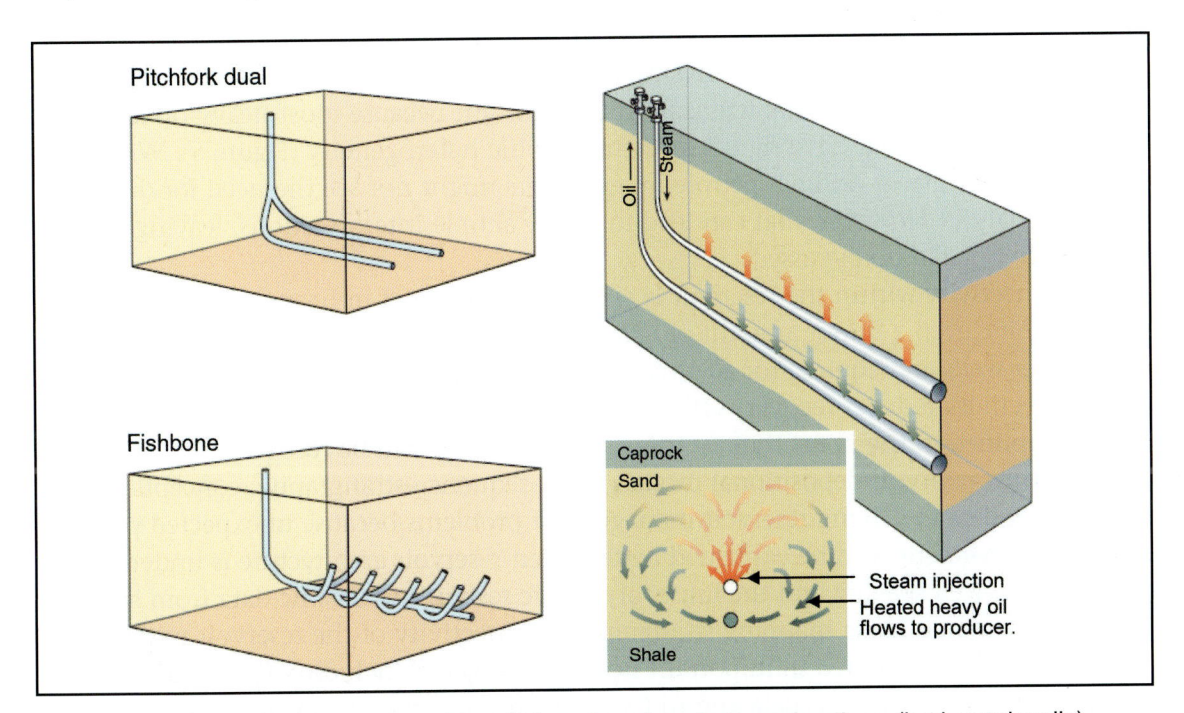

**Figure 3.** Heavy-oil production can benefit from long borehole perforations (horizontal wells). Depending on the geology or the viscosity of the oil, several multilateral drilling techniques can be used to maximize recovery. In steam-assisted gravity drainage, two horizontal wells are drilled in parallel, one for steam injection and the other for production (Curtis et al., 2002).

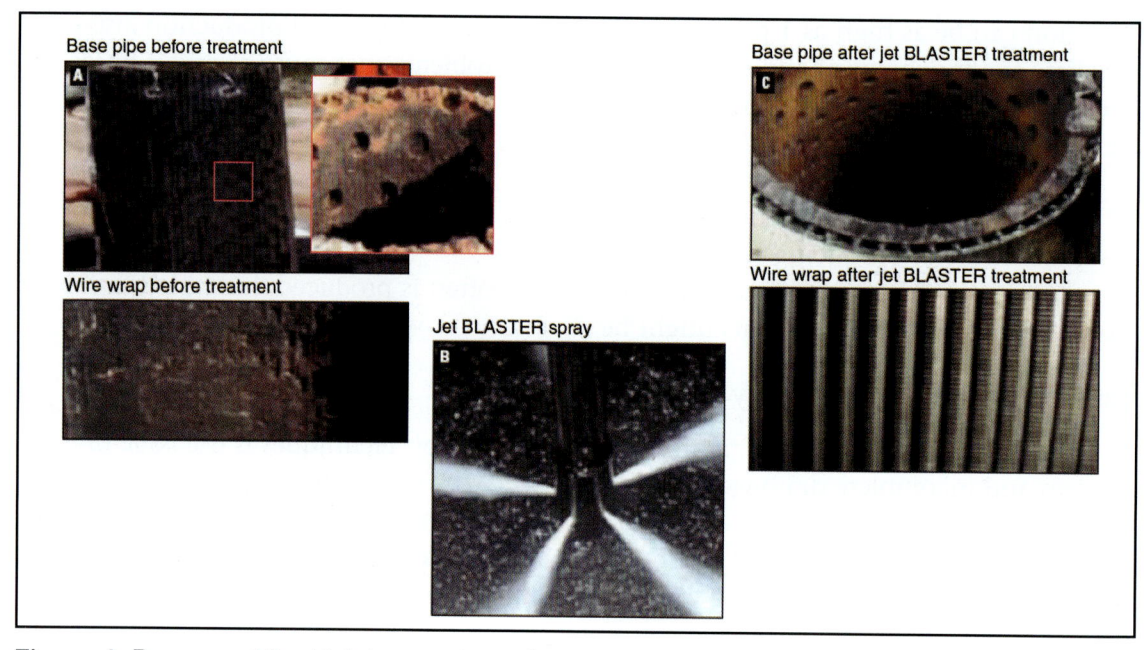

**Figure 4.** Because of the high temperature of common heavy-oil production environments, a great deal of thermochemical reaction causes wellbore plugging such as the scale on production pipe shown here. In addition, much reservoir sand can be produced along with oil, which requires screening and/or separation (Curtis et al., 2002).

## Geophysical applications

In heavy-oil production, geophysical activity is aimed at reservoir characterization and reservoir monitoring. Characterization is important because most heavy-oil fields are in clastic reservoirs that have significant stratigraphic heterogeneity (Figure 5). With the high spatial density of drilling, petroleum engineers might not see the need for detailed geologic analysis. However, that can lead to overly simple interpretations, leaving stratigraphically isolated oil bypassed.

Connectivity within the reservoir is especially important in heavy-oil recovery. In gas reservoirs, even a minor amount of connectivity allows gas to migrate to the producer, so fewer wells are required for adequate recovery. In light-oil reservoirs, the effect of low connectivity reduces recovery efficiency by as much as 50%. In heavy-oil reservoirs, that is amplified to the point that as little as 5% of the oil will be recovered.

Detailed reservoir depositional models using sequence-stratigraphic concepts can help a great deal in solving reservoir-management problems because unexpected pressure and fluid movements can be explained better when reservoir architecture is understood. Despite close well spacing and the limits of seismic resolution, seismic data from surface acquisition are helpful in mapping stratigraphic connectivity of the reservoir model. That has been documented in important heavy-oil projects, including the Hamaca field in the Orinoco belt of Venezuela and in Coalinga in California.

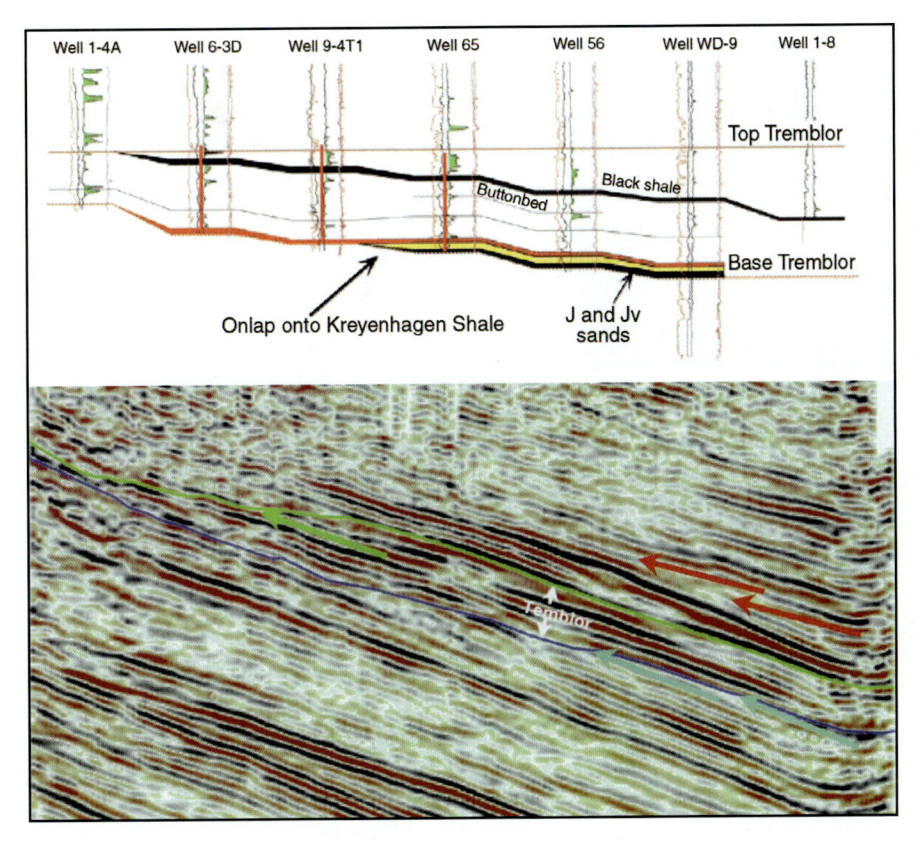

**Figure 5.** Seismic data can be important in reservoir characterization of heavy-oil fields. Stratigraphic compartments isolate steam injectors and production wells. The Coalinga heavy-oil field produces from the Temblor Sandstone. The reservoir overlies truncated Eocene strata (blue arrows) and underlies a surface of truncation (green) beneath Pliocene strata (onlap in red) (Clark et al., 2001).

## 4D seismic

In heavy-oil fields, surface seismic data also are used in time-lapse monitoring of reservoir production. With multiple seismic surveys over time, significant changes in the subsurface can be viewed during production, especially if steam or high heat is being introduced. Some seismic effects show good correlation with only one change in attribute of a reservoir (e.g., tracking steam). More often, the effects are complicated and require consideration of the whole dynamics of the reservoir because saturations, pressures, and temperatures change.

## Duri 4D seismic example

The Duri steam-flood monitoring in Indonesia is a good example of the situations described above. When steam was injected into the central well, it was anticipated that it would be a year or more before heating effects would reach observation wells or producers. Four-dimensional seismic was used to monitor the heated reservoir by measuring time delays. The estimate of time for heat to reach the observation wells was approximately correct. However, another effect was noted much sooner. Within a few months, the pressure wave from steam injection advanced throughout the reservoir. The wave of increased pressure caused gas in the reservoir to return to solution, which increased the velocity of the reservoir interval. That was detected by traveltime advances below the reservoir in 4D seismic monitoring.

Forward seismic modeling of those conditions successfully matched seismic observations (Figure 6). The horizon time-difference maps show the pressure wave in the first

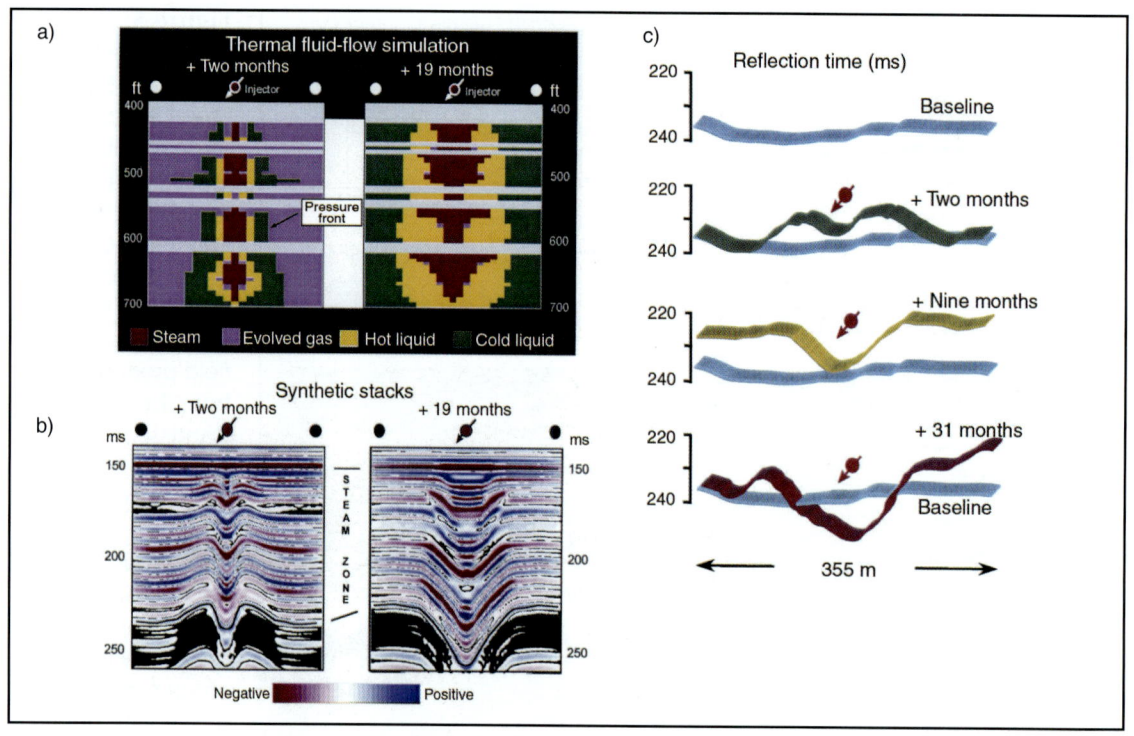

Figure 6. Duri field, Indonesia. Reservoir monitoring of heavy-oil fields must take into account all aspects of the reaction of the subsurface to injection. Here, injection of steam at first pushed gas back into solution and created higher velocity and a time pull-up of the reservoir. Later, when the steam expanded, higher temperatures "melted" the oil, causing it to reverse and become lower in velocity. (a) The reservoir model is used to compute (b) the theoretical geophysical response. When actual seismic data were acquired, (c) measurements from the base of the steamed zone show those effects (Jenkins et al., 1997).

five months, followed by the push-down from steam (Figure 7). When several surveys were taken over part of the field, they revealed heterogeneity in the spreading of steam because of the stratigraphic complexity of the reservoir (Figure 8). From such information, better reservoir management can be performed with intervention techniques such as huff and puff (injecting steam and producing it back with oil), steam isolation (shutting in the producer), choke installation (slowing the injection), and injector shut-in (stopping the injection).

Both time-delay information and amplitude effects were accounted for in the reservoir response to time-lapse seismic data. Therefore, the 4D data were used as a volumetric-pressure transient test in the early months of injection and later were used to illustrate the advance of the viscosity front between wells.

## Borehole seismic data

In addition to surface seismic data, access to shallow boreholes allows application of downhole geophysics. Signals from sources at the surface can be recorded in the borehole (or vice versa), and a detailed view of the reservoir interval can be obtained in a vertical seismic profile (VSP) (Figure 9). VSP images a cone-capped cylinder of the subsurface with the potential for higher resolution than that of the surface seismic data by elimination of the upgoing raypath through the shallow subsurface. With recently improved instrumentation and processing, more 3D VSP images are being used for characterizing and monitoring reservoirs for higher resolution of geology and reservoir dynamics near the borehole.

Boreholes also can be used to acquire seismic data downhole from one well to another in a crosswell geometry. Special downhole seismic sources have high frequency and good penetration without harming the borehole. Data received at the borehole where geophones have been deployed are analyzed for traveltimes (tomography), and the reservoir interval is mapped for velocity. Crosswell seismic data also have been processed for reflection response and show very high resolution of reservoir geology (Figure 10). Crosswell seismic also can be acquired in time-lapse mode for reservoir monitoring of production effects.

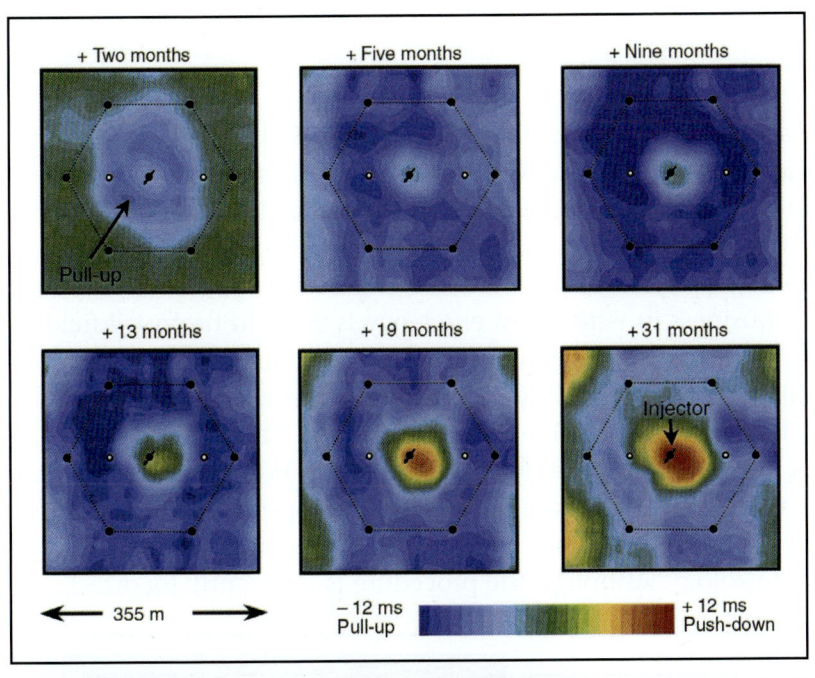

Figure 7. Duri field, Indonesia. Maps of time delays observed at the base of the steamed zone show the initial pull-up from the pressure wave, with evidence of gas going into solution. That was followed by push-down caused by the advance of the steam front (Jenkins et al., 1997).

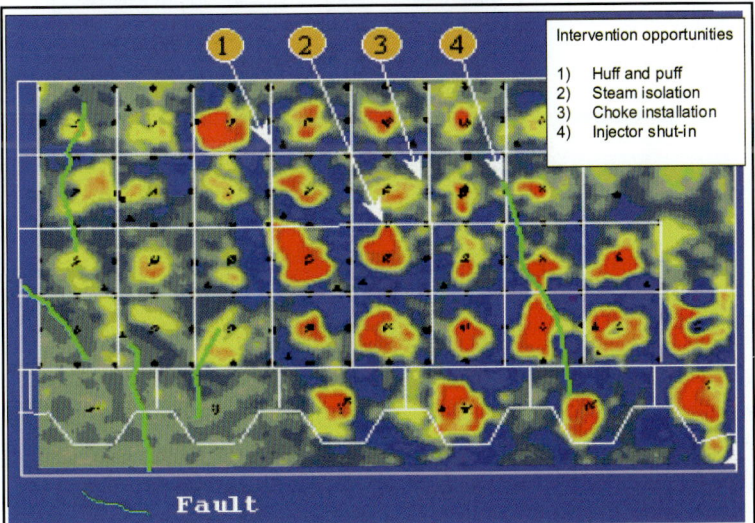

Figure 8. Duri field, Indonesia. Four-dimensional seismic data were acquired over a substantial area in the Duri field. High value in the energy-amplitude maps shows the uneven sweep of viscosity fronts. Intervention options for obtaining a more uniform sweep include (1) huff-and-puff steam injection, (2) steam isolation, (3) choke installation, and (4) injector shut-in (Sigit et al., 1999).

## Crosswell electromagnetics

A method less commonly used than seismic data, an electromagnetic source, also can be used downhole in fiberglass-cased observation wells in a crosswell geometry (Figure 11). Measurements can be inverted for definition of reservoir properties such as water saturation, porosity, and clay content. Electromagnetic data are especially powerful in detection of the presence of steam, anticipating steam breakthrough, and locating bypassed oil. High temperatures associated with steam injection can cause formation resistivity to decline by as much as 40%. Measurements of resistivity can be inverted to produce saturation images between wells at approximately the same resolution as the seismic tomography. Together, seismic and electromagnetic images provide the best estimates of reservoir continuity, volume, and heat distribution between wells (Figure 12). Those are some of the most important contributions geophysical data can make to assist heavy-oil field production.

## Remarks

Combined, the applications of geophysics for heavy oil are used to add value to the project by (1) evaluating the total resource size through reservoir characterization, (2) maintaining production rates by monitoring drainage, and (3) contributing to efficient heat management by monitoring distribution of steam. But what price should be paid for this technology? Heavy-oil geophysics is an additional expense that adds helpful but perhaps marginal value in a business in which cost efficiency is vital. In heavy-oil fields, wells are spaced closely, and geology is sampled densely. Moreover, substantial investments already exist in monitoring of heavy-oil reservoirs by instrumenting downhole injectors and producers and by drilling specialty observation wells. Therefore, there are cases in which incremental information gained from geophysics is not cost-effective.

Production of heavy oil is almost never simple. Many factors can contribute to unexpected problems. Many of those factors are mechanical or chemical, having almost nothing to do with interwell geology. Although the procedure is inherently incorrect,

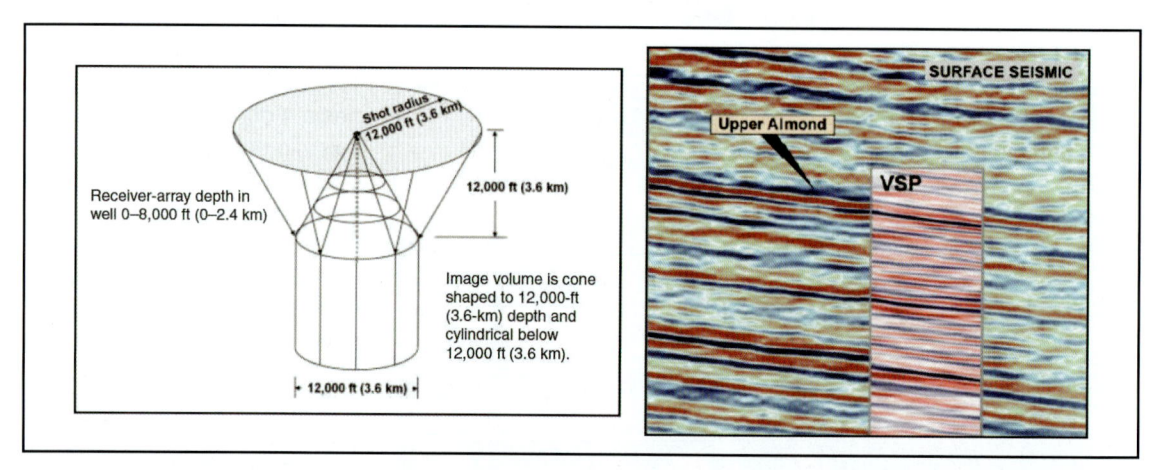

Figure 9. Vertical seismic profiling (VSP) is an acquisition in a borehole from multiple surface sources. This results in a cone-and-cylinder image around the well. The downhole seismic data potentially show considerably higher resolution than do surface seismic data (Paulsson et al., 2004).

modeling of heavy-oil fields often is done by regarding the reservoirs as homogeneous slabs. Thus, making a successful case for investment in reservoir geophysics for heavy-oil fields requires an approach that (1) supports geologists in their need for predicting the effects of reservoir heterogeneity and (2) provides a value-of-information (VOI) explanation to managing operators (Figure 13). When operators of the field become aware that the benefits of geophysics exceed the costs, they become supporters of reservoir-geophysics technology in the production of heavy oil.

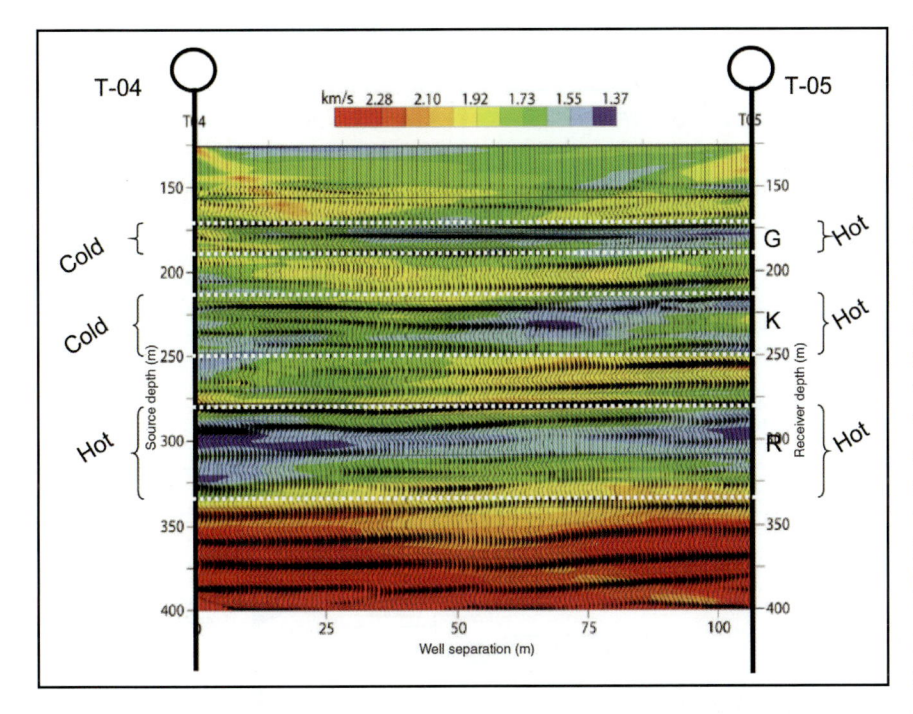

Figure 10. Crosswell seismic data are obtained with downhole sources in one well and receivers in another. The potential exists to obtain high-resolution reflection data as well as direct measurements of velocity. Here, low velocity shows that steam advance from a nearby injector has reached well T-05 in intervals G, K, and R but has reached well T-04 only in interval R (Hoversten et al., 2004).

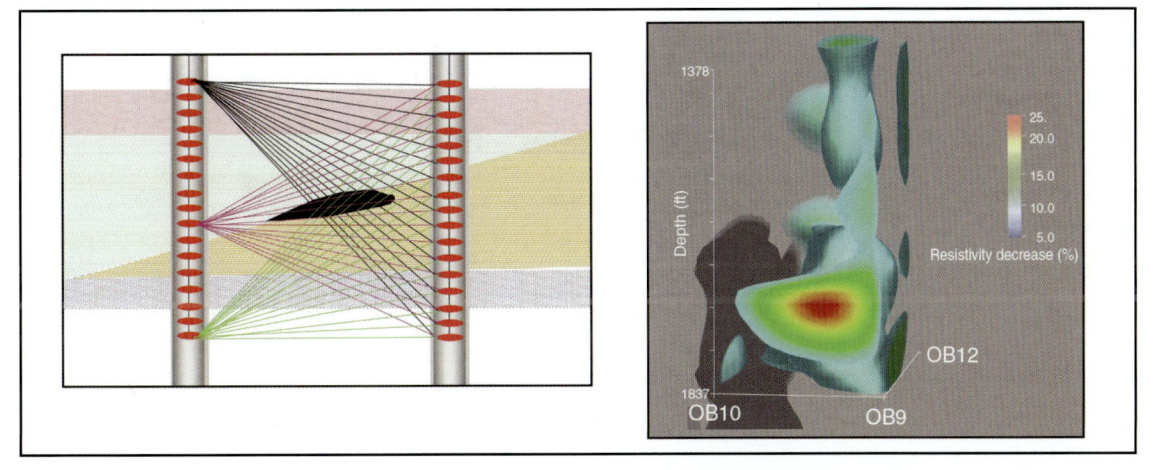

Figure 11. Crosswell electromagnetics also can be obtained in fiberglass casing to acquire a tomographic view of the steamed zone. If several wells are used simultaneously, a 3D view is obtained (Wilt and Morea, 2004).

**Figure 12.** Comparison of the crosswell tomogram of velocity from seismic measurements with the crosswell tomogram of electromagnetics shows general agreement on reservoir characteristics (Hoversten et al., 2004).

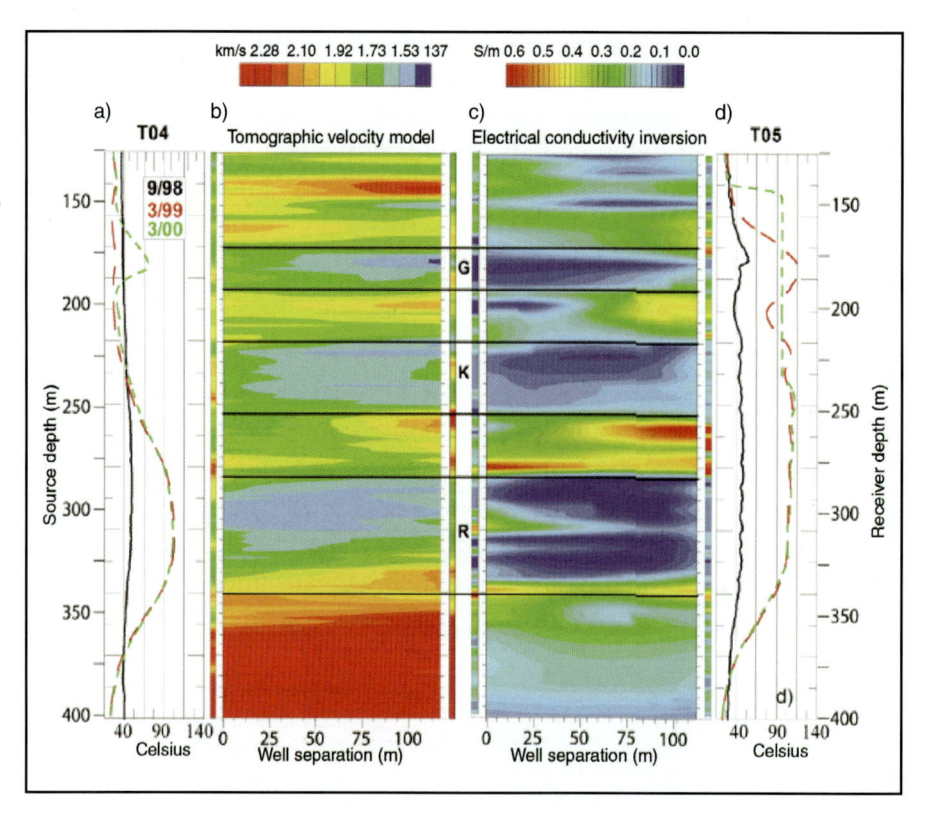

**Figure 13.** A value-of-information (VOI) example for heavy-oil geophysics. Because it is often easier and cheaper to drill wells in shallow heavy-oil fields, the value of expensive geophysical work comes under greater challenge.

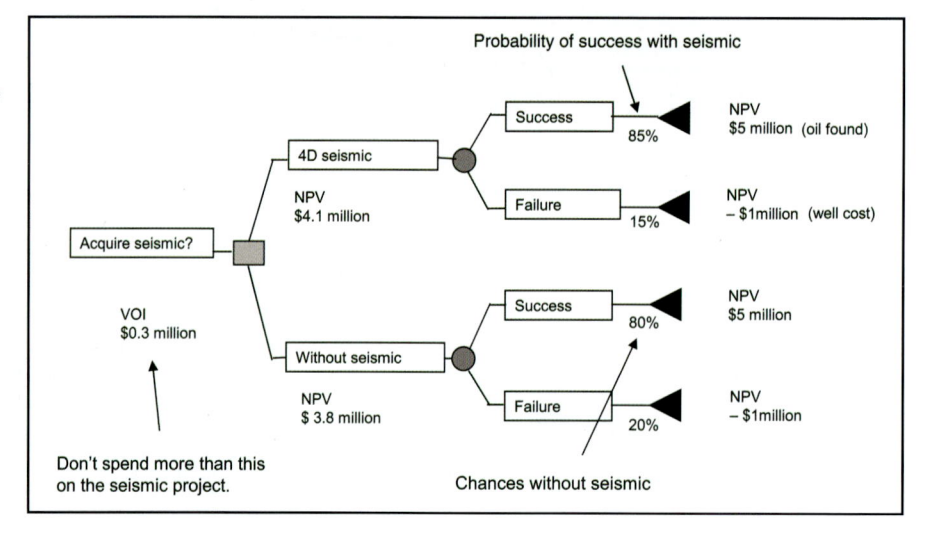

# Chapter 8   Reservoir Geophysics in Carbonates

This section of the book is intended to provide familiarity with geophysics in carbonates. Much of the historical effort for geophysical applications in reservoirs is based on the use of seismic data in clastics (sandstone). Application of seismic data in reservoir work for carbonates has lagged behind that for clastics for at least three unrelated reasons. First, the larger carbonate reservoirs are onshore in areas where near-surface conditions are difficult for obtaining quality seismic data. (That fact will not be addressed in this text, but the following two points will be.) Second, the geometry of carbonates is more difficult to predict than that of clastics because of complex original depositional facies and the impact of diagenesis. Third, the use of acoustic properties of carbonates for detection of fluids and reservoir-property discrimination is challenging.

**Will reservoir geophysics work in carbonate fields?**

With the advent of 3D seismic data, reservoir geophysics advanced significantly in noise-free seismic acquisition and favorable reservoir acoustic response. In general, those advances have occurred in mapping of offshore fields in clastic reservoirs, where sands are separable from shales in seismic response. Many excellent and even spectacular case histories show how seismic data can depict structure, depositional environment, reservoir properties, fluid content, hydrocarbon saturation, and the dynamics of pressure and fluid movement during production.

But what about carbonates? Because the largest light-oil reserves in the world are in carbonate reservoirs, it is important to determine how many of the successful reservoir-geophysics applications in clastic rocks are applicable to carbonates. Commonly, geoscientists believe that carbonates are so high in velocity and so "stiff" that reservoir facies or fluids cannot be determined directly — e.g., there are no bright spots to find, and amplitude maps do not work. That assessment might be true for exploration, but it is a false assessment of the proven application of reservoir geophysics in carbonates.

**What properties differentiate carbonate reservoirs?**

Carbonates are different from clastics because they are bioconstructed and clasts in them have not been transported very far (Figure 1). The matrix of the rock might not have been very worn mechanically and can retain fairly sharp and angular faces. Organic construction works when organisms precipitate calcium from clear and warm seawater. The organisms range in size from very small (microscopic) to rather large (larger than a hand sample). Because the organisms grow and die, they leave residual carbonate shells, supports, or secretions.

Carbonate platforms generally are classified by geometric relation to the shore and sea level. Categories include rimmed shelves, ramps, epeiric platforms, isolated platforms, or drowned platforms (Figure 2).

**Figure 1.** An aerial photograph of a modern carbonate deposit shows reef and back-reef environments and tidal channels.

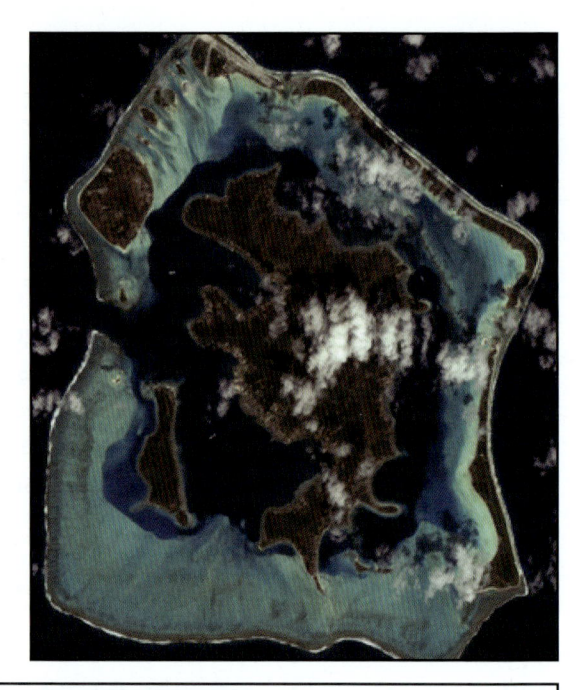

**Figure 2.** Carbonate platforms are classified as general concepts of shape in relation to water depth, shelf margin, and shelf break. In a given area, it might be possible to find more than one platform type, and they might grade into one another.

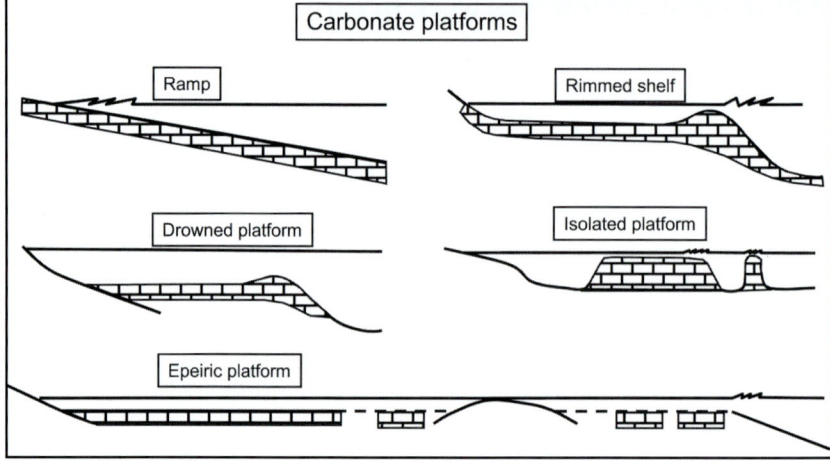

As the underlying basement (or sea level) rises or sinks, facies are exposed subaerially or shift positions laterally or the system is drowned and carbonate production shuts down. In addition, nonreservoir facies interfinger with clastics (sand or shale) in humid environments or with evaporates in arid environments.

The original facies distribution of reservoir quality has the tendency for significant horizontal and vertical heterogeneity over short distances. That is confirmed by observing some modern analogs (Figure 3). In contrast, epeiric platforms were deposited over extremely large areas with very low gradient and have the potential for properties with large correlation lengths. There are no modern analogs of such platforms.

## Seismic facies mapping

Geophysics has been shown to be a powerful tool for predicting stratigraphic characteristics of clastic reservoirs. At the exploration and development scale, seismic data

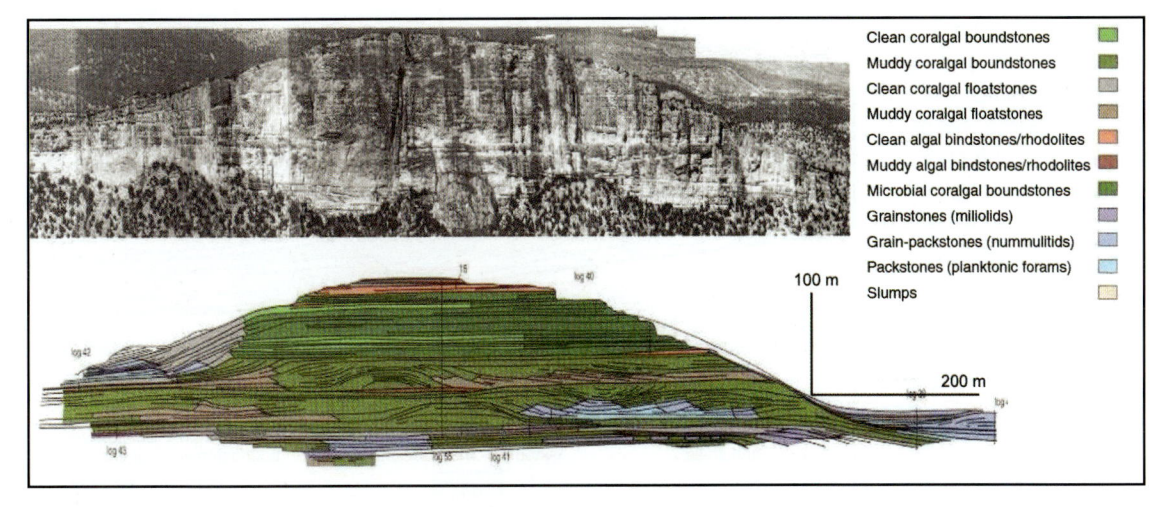

Figure 3. Detailed study of a Miocene isolated platform in Mut Basin, Turkey, shows how abruptly rock types differ horizontally and vertically. At some localities, reservoir properties of carbonate rocks can have exceedingly short correlation distances (tens of meters) (Bassant et al., 2004).

also contribute to understanding of carbonate reservoirs by use of similar principles of sequence stratigraphy. The special nature of carbonate-sediment deposition is taken into account. That requires an appreciation of extensive shoals, belts, and margins and the constructional nature of reefs and mounds. Constructional carbonate deposits attempt to maintain an elevation close to sea level.

The first step in facies prediction is to determine the systems tracts bounded by unconformities. In carbonate-rock terrane, that process can be confusing when multiple erosional surfaces converge on top of constructed carbonate bodies. In addition, because carbonate fragments often bind during construction, they can have locally steep dips where several reflections can converge (Figure 4). However, a facies prediction can be made from seismic data when sequence boundaries are determined, stratal patterns of onlap and downlap are accounted for, and internal reflections are accounted for (Figure 5).

When carbonate facies are the major factor in predicting reservoir quality of the field, seismic data become a powerful tool, but with limitations. Like seismic data of clastics, reflections from the reservoir are limited by vertical resolution. However, in carbonates, that shortcoming is exaggerated by the generally faster velocity of carbonate minerals. This lowers the frequency content (and thus vertical resolution) of the data. In comparison with clastic rocks, many carbonate rocks have horizontal stratigraphic differences but limited or confusing variation in reflection character. Therefore, facies boundaries are less well resolved and are harder to recognize in carbonate rock.

An example of how complex seismic response can be is illustrated in the interpretation of facies from platform carbonates (Figure 6). In an otherwise fairly flat terrain, seismic reflectors representing carbonate facies changes show significant variation in dip rate and reflection amplitudes. Knowledge of carbonate geology and seismic-interpretation experience are important in successfully mapping carbonate facies. The more time one spends in interpretation, the better one becomes at determining the important characteristics of that special type of seismic data.

**Figure 4.** A schematic diagram shows the relationship among source, reservoir, and seal facies for different platform types. Source, reservoir, and seal facies often are found in an arrangement suitable for trapping hydrocarbons (Handford, 1998).

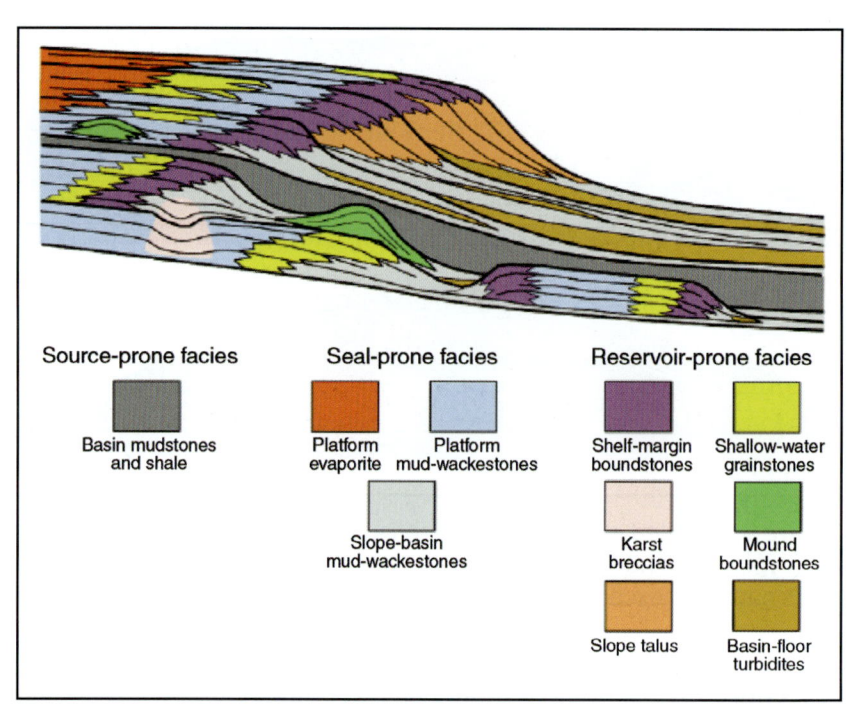

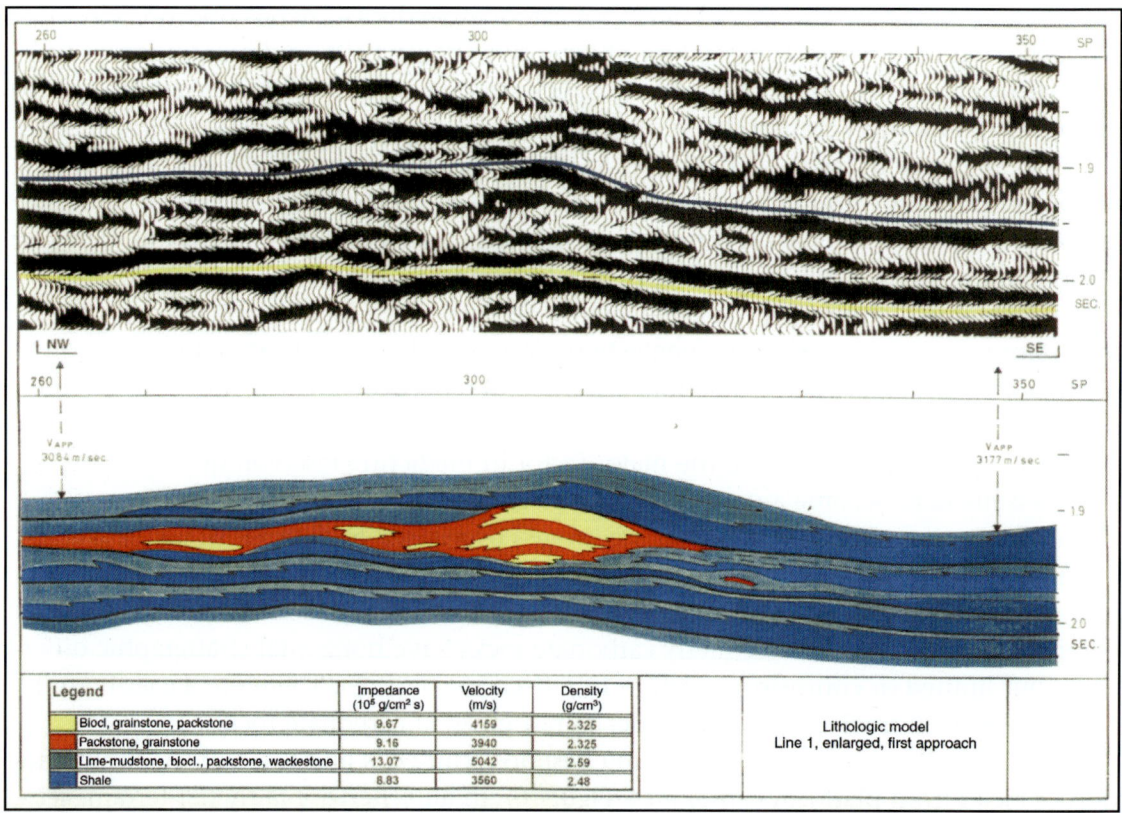

**Figure 5.** Interpretation of seismic data from carbonate rocks requires using a conceptual model of carbonate systems. Here, facies are interpreted with reservoir possibilities shown in red and yellow (Fischer et al., 1997).

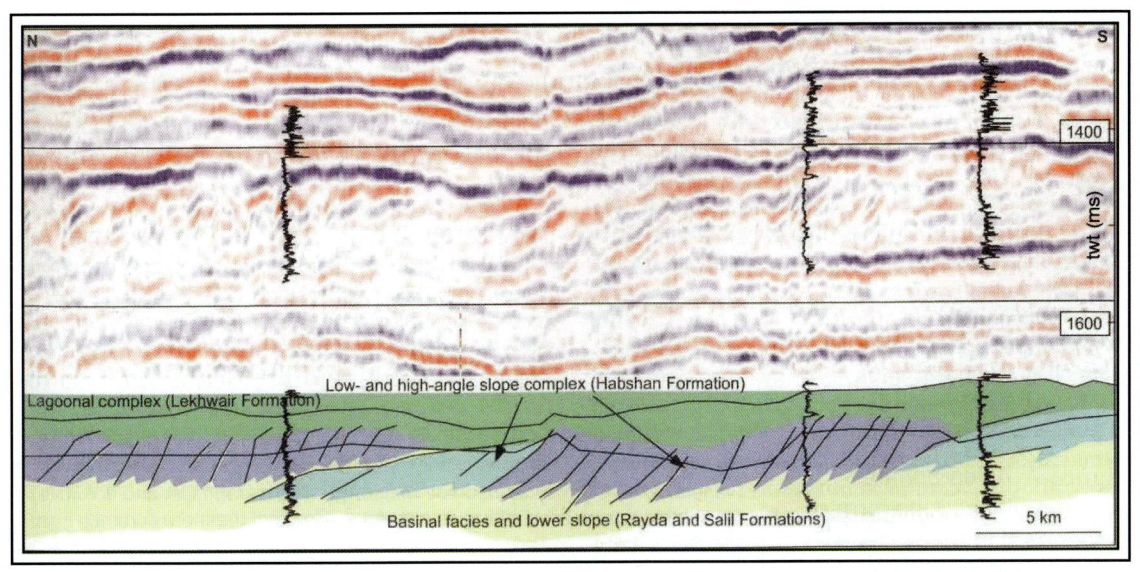

**Figure 6.** In epeiric carbonates, facies are spread out over long distances and might not all be present in any one place. However, complex reflections internal to the carbonate system can represent very local depositional changes related to depositional tracts (Droste and Van Steenwinkle, 2004).

## Diagenesis

The generally complex rock fabric resulting from diagenesis — chemical alteration of the reservoir — also contributes significantly to heterogeneity in carbonates. Several types of diagenesis can occur in carbonates (Figure 7). The secondary minerals calcium, aragonite, and dolomite are precipitated or dissolved with small changes in pressure, temperature, pH, and concentration of $CO_2$. Therefore, pore shape, size, and connectivity of the carbonate reservoir are fashioned by the combination of the original depositional fabric, the burial history, and diagenesis. Work on seismic-reflection continuity in carbonate reef systems shows that mapped horizons do not necessarily follow original depositional facies but are influenced strongly by interfaces resulting from diagenesis. Thus, the internal reflection systems that are used to understand sequence stratigraphy of carbonates can be thought of as mapped diagenetic boundaries rather than original depositional facies (Figure 8). That fact is important to know when trying to sort out facies and diagenesis from seismic reflections.

In a case history of the Malampaya deepwater carbonate field (Fournier, 2007), five major diagenetic events are proposed. The first was the common marine diagenesis of calcite that occurs during and immediately after deposition. The next two events were results of freshwater chemical interactions (meteoric). The final two events occurred during burial (Figure 9). As a consequence, the section of rock that had the highest porosity now has the lowest porosity. A view of the seismic data for the whole field (Figure 10) suggests that the extended internal reflections of the central interior carbonate platform are related not only to original depositional facies but also are influenced strongly by diagenesis.

## Reservoir characterization

Commonly, reservoir models made from interpolating well data alone are too simple to predict reservoir properties accurately. Carbonate-rock reservoirs are no exception. If seismic data can help reduce uncertainty about rock properties among wells, it often provides high value in reservoir management.

Volumetric characterization of carbonate reservoirs must include a thickness estimate, a conceptual geologic model (facies relations), estimated locations of facies boundaries, and a description of a relation between facies and effective pore space. In carbonates, that information includes mapping of diagenetic and depositional facies. When tied to well data, seismic character can be used to estimate reservoir presence and quality between wells.

**Figure 7.** All carbonates undergo at least one form of diagenesis. In many cases, several periods of chemical alteration occur, and they are not always localized to occur within the original depositional facies.

Diagenesis — Carbonate reservoir minerals (calcite, aragonite, and dolomite) are subject to significant chemical alteration. Some of the more common alterations are:

- marine seawater cementation with some porosity retention
- meteoric groundwater dissolution (or cementation)
- dissolution from subaerial exposures
- burial and cementation from connate waters
- dissolution from vertical flow of hydrothermal waters
- dolomitization (replacement) from Mg-rich waters

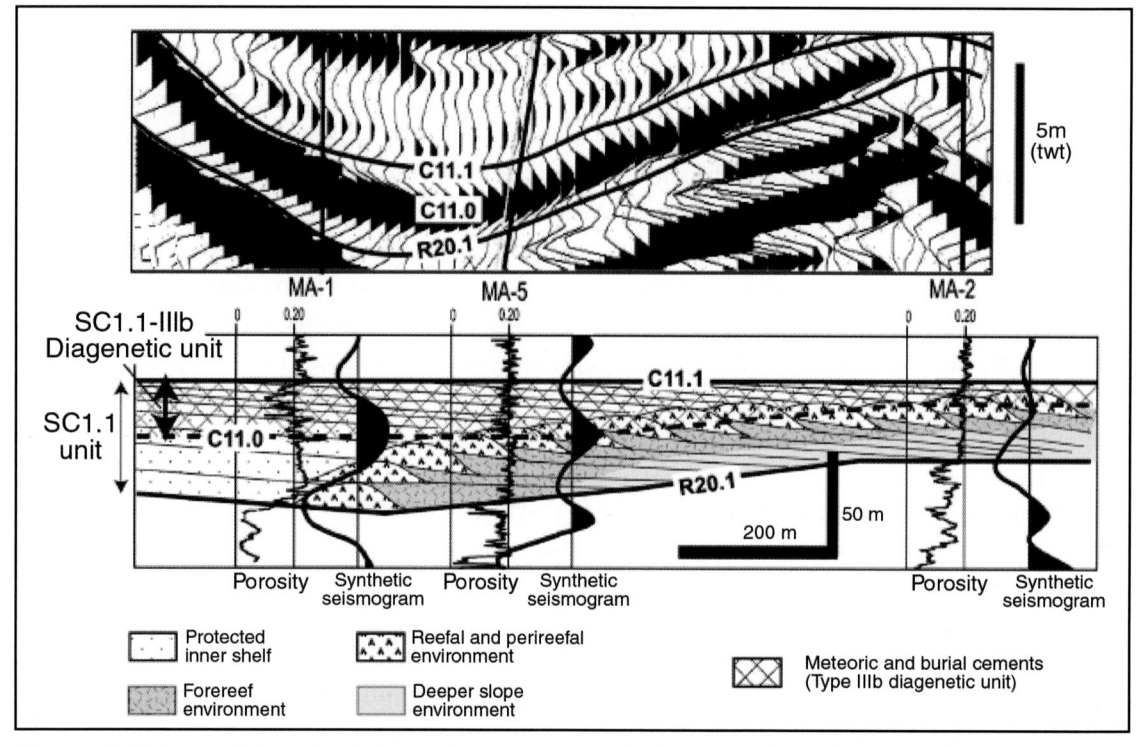

**Figure 8.** This careful study of the carbonate seismic facies from deposition and diagenesis at Malampaya shows that reflections do not necessarily follow original depositional time lines or facies. Some follow the diagenetic overprint instead (Fournier and Borgomano, 2007).

Even in carbonates, there are circumstances in which the information critical for reservoir management can be correlated with a single seismic attribute. For example, in the Ghawar field (Figure 11), the Kuff-C producing interval is associated with low seismic impedance, and a successful development strategy was used to exploit that relationship (Dasgupta et al., 2001).

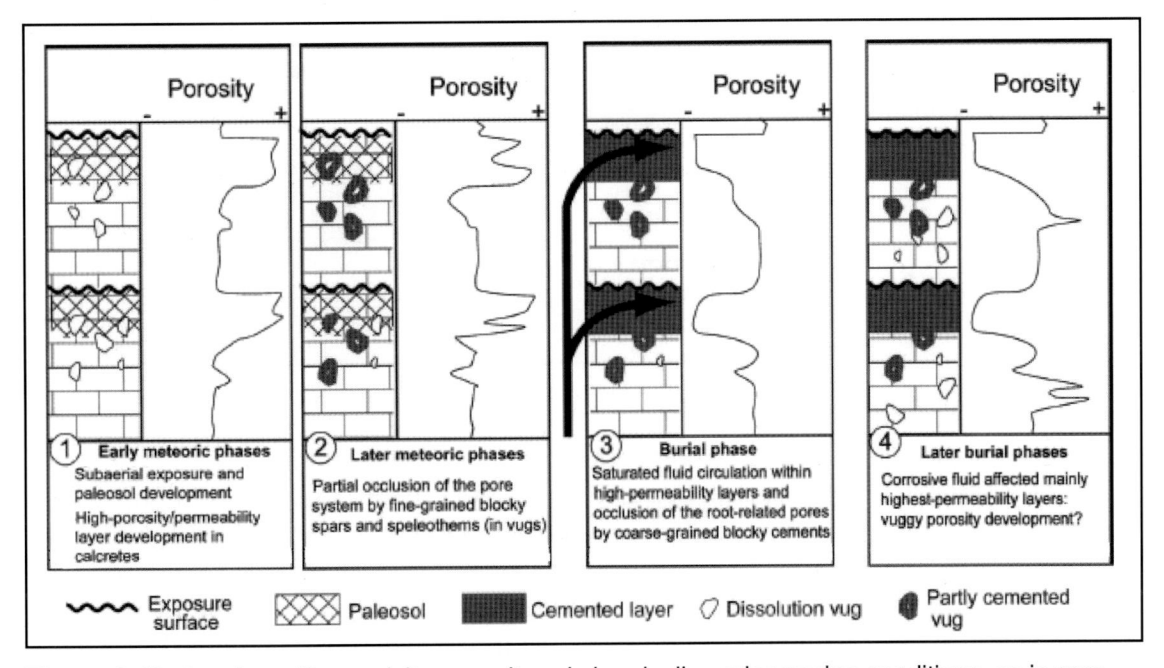

Figure 9. During deposition, calcite was altered chemically under marine conditions, as is common in most carbonates. That lowered the initial depositional porosity. The subsequent diagenetic history of the Malampaya field also shows four additional periods of diagenesis, during which significant porosity was formed and destroyed and additional porosity was developed (Fournier and Borgomano, 2007).

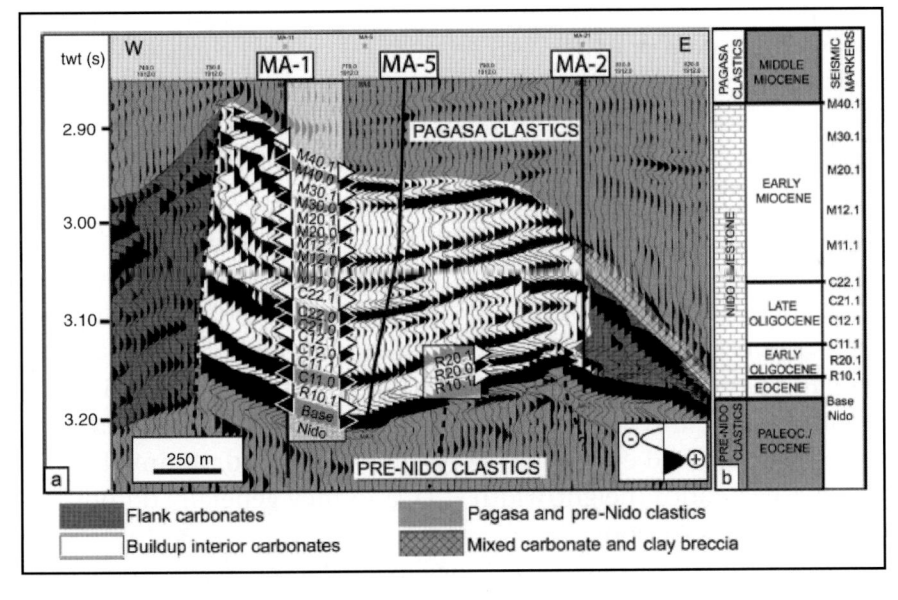

Figure 10. A seismic cross-section view of the entire Malampaya field shows that the central carbonate platform has multiple parallel and continuous reflectors that might be the result of diagenetic events (Fournier and Borgomano, 2007).

**Figure 11.** In the Kuff-C carbonate reservoir of the Ghawar field in Saudi Arabia, porosity can be related to seismic amplitude response. The inverted data were used to determine a successful round of additional development drilling (Dasgupta et al., 2001).

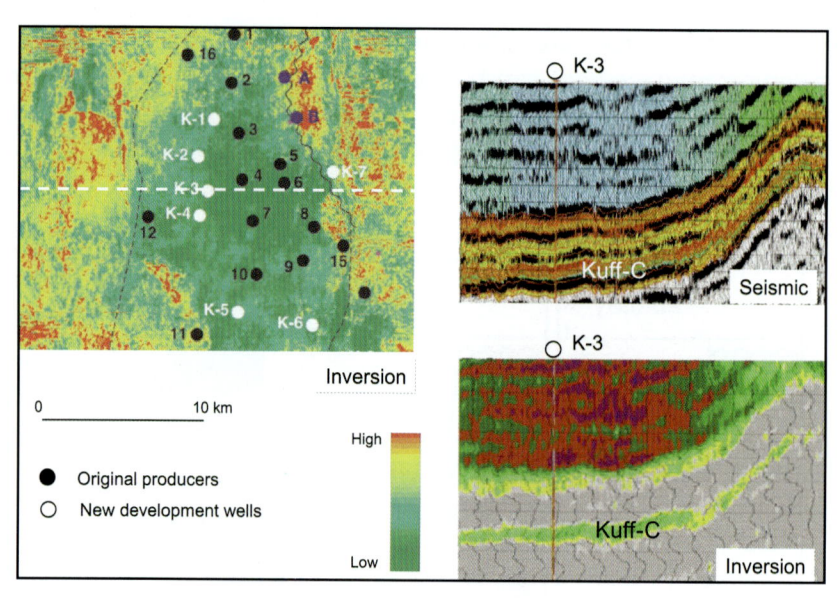

## Reservoir properties — Special considerations

Geoscientists who practice carbonate-reservoir prediction commonly are frustrated that they cannot successfully predict reservoir properties before a significant number of wells are drilled. Principles that appear to predict good reservoir in one field do not apply in another because of differences in diagenetic history. If geophysics can help make useful predictions prior to drilling, it is of much value in reservoir management.

How difficult is it to predict carbonate-reservoir parameters from seismic data? Certainly, it is not easy. Seismic-amplitude applications in clastic reservoirs have not made a direct transfer to exploitation of carbonate reservoirs. Parameters of clastic reservoirs are easier to predict from seismic data; the rocks generally are much more compressible than carbonates (Figure 12), primarily because of clasts of shale between sand grains (Figure 12).

Carbonate minerals tend to be welded in a "stiff" fabric. Petrophysical modeling of carbonates shows that those "hard" rocks do not change much in acoustic character when porosity varies only moderately or when petroleum has displaced connate waters. That is the logical extension to carbonate rocks of the well-documented acoustic theory of seismic modeling in clastics (i.e., the Gassmann theory). It has discouraged workers for years from trying to measure reservoir attributes from seismic data in carbonate rocks. However, the underlying assumptions of the Gassmann theory might be violated in carbonate rocks.

Petrophysical work on carbonate seismology has progressed and shows promise when pore shapes are taken into account. Velocity in carbonates varies from slow to very fast, depending on whether pore spaces are supported by grain or framework. The acoustic response of carbonate rocks is dependent on pore shape. Pores that are predominantly spherical are supported strongly by a well-connected and well-cemented matrix where stress from acoustic waves is distributed evenly and pores are compressed minimally. Those are pores with high-aspect ratios (height/width = 1), and they generally do not permit the direct detection of hydrocarbons.

By contrast, low-aspect-ratio pores are more elliptical or "flat," and during the passing of seismic waves, the matrix does not support the pore nearly as well. Shales and

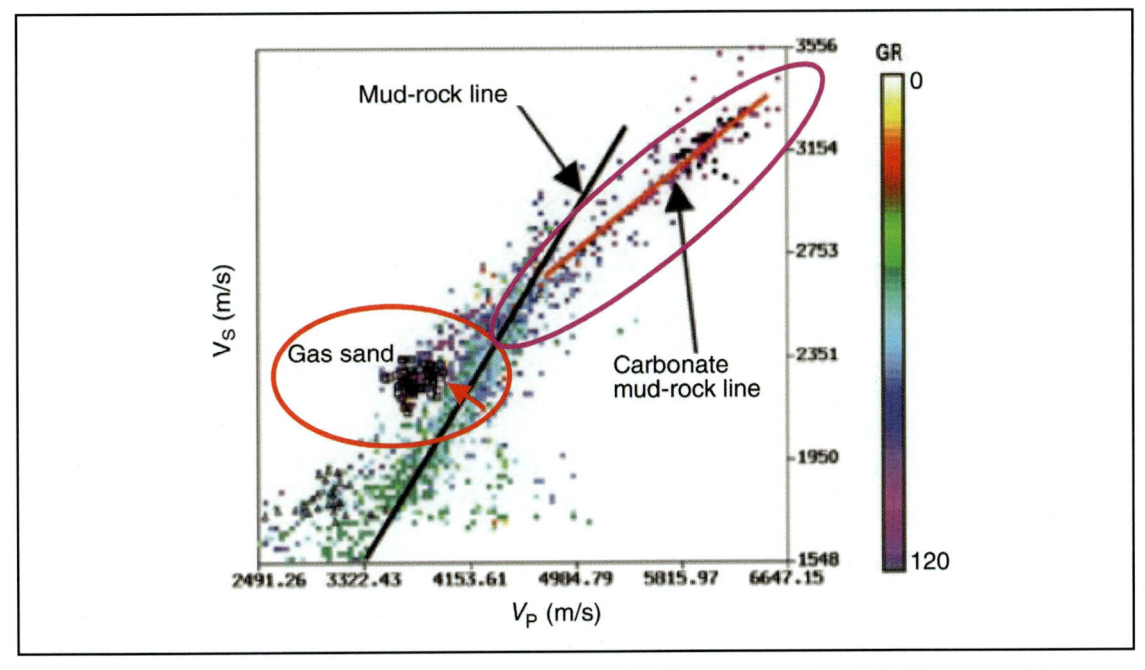

**Figure 12.** Carbonate rocks and clastic rocks show significant petrophysical differences. These data from wells in the Canadian foothills illustrate the high velocity of carbonate reservoirs compared with velocities of sandstones of the same age. The compressional ($V_P$) and shear ($V_S$) waves are crossplotted so that the trends can be understood better. Note that the mud-rock line for clastics is a lithology trend from which gas sands deviate (red arrow). A separate mud-rock line exists for carbonates, and the search has been made to find the equivalent fluid-indicator separation in $V_P/V_S$ space. Unfortunately, rock-physics concepts that are useful in clastics cannot be used directly for evaluation of carbonate rocks (Li et al., 2003).

fractures also contribute to the lowering of the pore aspect ratio. Such pores have good potential for response to differences in fluid and pressure content and hold promise for direct hydrocarbon detection or monitoring of production (Figure 13).

## Reservoir monitoring

The difficulties in measuring fluid and pressure differences in carbonate rocks suggested by the Gassmann theory do not mean reservoir monitoring is impossible. Where low-aspect-ratio pores exist, they consistently have demonstrated a better seismic response than expected from classical petrophysical analysis. In comparison with clastics, carbonate rocks show more factors that affect pore size, shape, and connectivity. Carbonate pore type is a function of both the original depositional fabric, which varies a great deal, and of diagenesis, which adds significant complexity.

With discrimination by pore type, geophysics can be used to map out carbonate flow units or to monitor production in rocks with low-aspect-ratio pores. An example from Canada illustrates just such a case. Ng et al. (2005) describe the monitoring of a gas and solvent injection in a carbonate oil field. Injected solvents were detected through seismic monitoring. In a comparison of a base seismic survey with a repeat seismic survey, subtle time differences were detected (Figure 14). Carbonate facies with vuggy porosity and fractures are of lower-velocity response than average because of the low-aspect ratio of the pores. Facies with intergranular porosity that had pores of high-aspect ratio

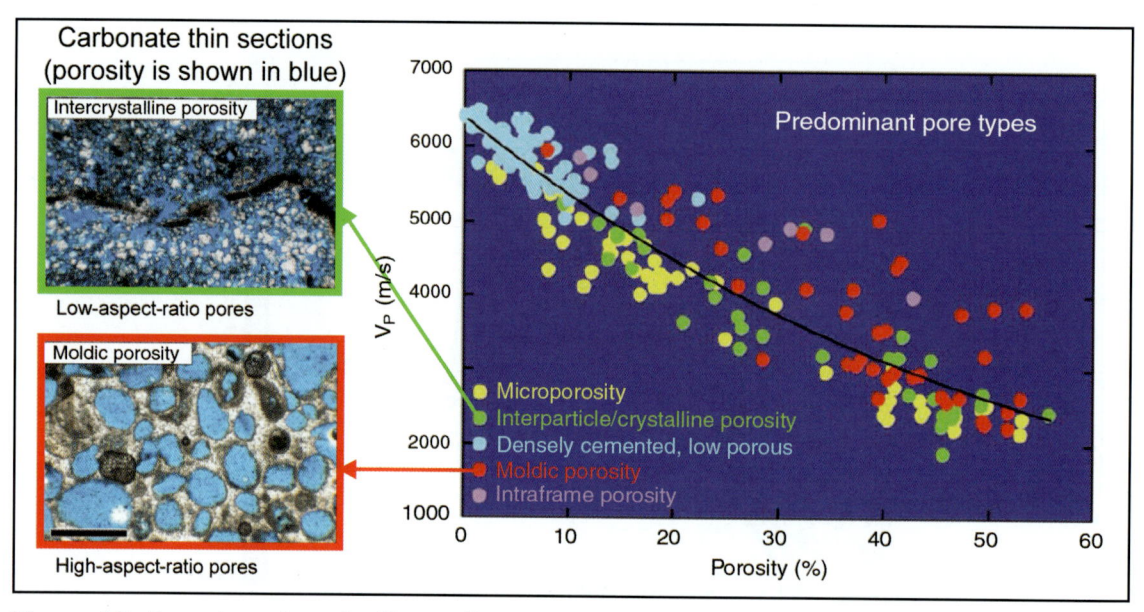

**Figure 13.** Pore shape has significant effect on acoustic seismic velocity. Rocks with high-aspect-ratio pores (red and purple) have higher velocity and are less compressible. Low-aspect-ratio pores (yellow and green) have lower velocity and respond more to fluid replacement. Rocks with these type of pores are good candidates for seismic monitoring (Eberli et al., 2003).

(rounder) did not respond, even though gas and solvent are known to have been injected in significant quantity (Figure 15).

## Fractures

Production from fractured reservoirs requires special management. In the early phase of production, loss of pressure in a very short period is normal because fractures are connected well. Primary production declines quickly. To maintain pressure, secondary recovery with downdip water injection is initiated early in the production history. Locations and fluid-flow rates of injectors and producers must be managed so that water does not break through fractures early and bring water production to unacceptable levels. Recently, injections of $CO_2$ and solvent have been used in tertiary-recovery methods, in which knowledge of fracture connections is also important for success.

Carbonate fields normally have some level of production from fractures. Carbonate rocks tend to be more fractured than clastic rocks. Carbonate minerals and the cemented matrix are brittle. In addition, because carbonate sediments are lithified early, the rocks are subject to fracturing throughout the formation of the reservoir and trap. Fractures are primary channels for movement of fluid during diagenesis. Thus, fractures are subject to leaching, mineral replacement, and cementation. Commonly, old fractures are cemented, and new ones are open and connected. It commonly is believed that present-day fracture orientation is predominantly vertical in most petroleum fields and is aligned with the direction of maximum present-day horizontal stress.

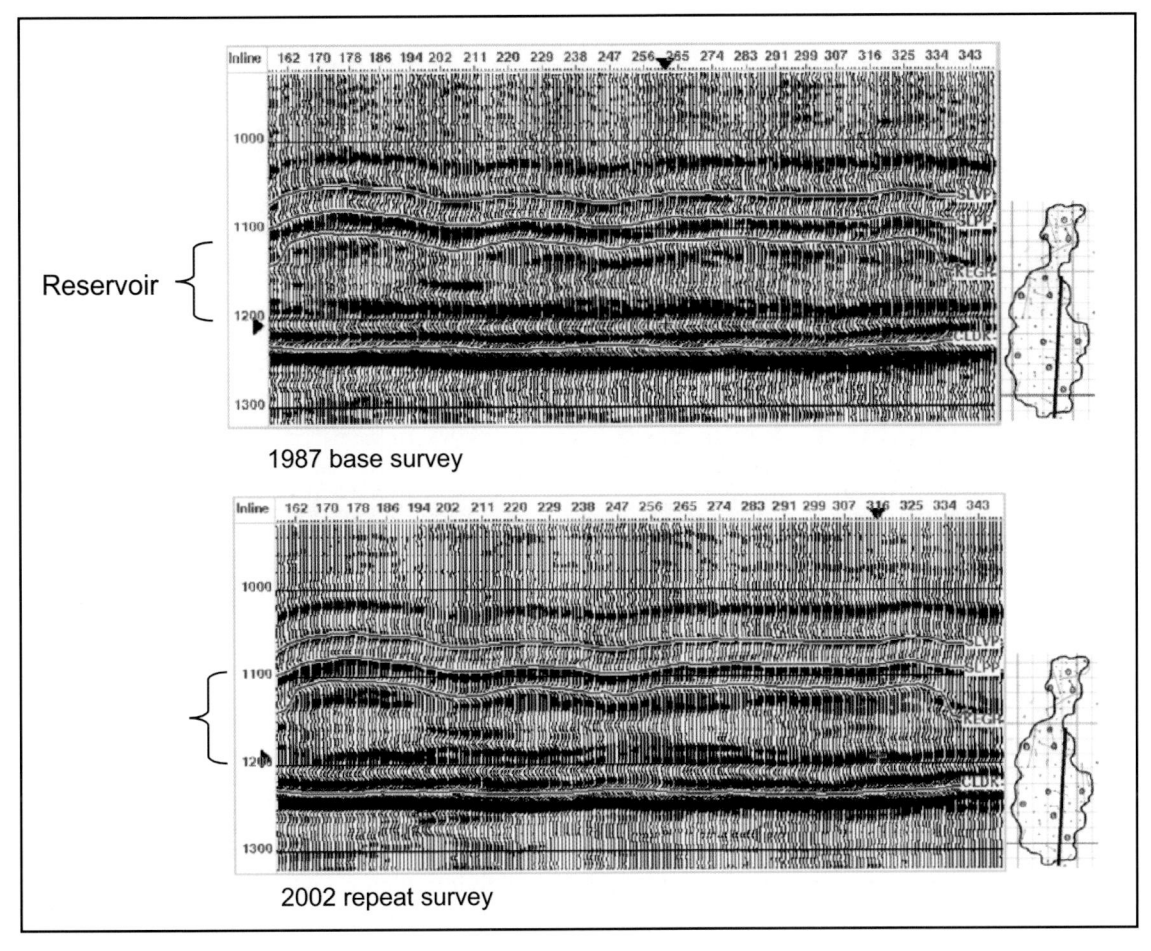

**Figure 14.** Seismic monitoring of gas and solvent in a carbonate reservoir in western Canada. These seismic cross sections are from base and monitor surveys. A traveltime delay was measured from reflectors above and below the reservoir where injected gas and solvent have replaced oil (Ng et al., 2005).

Reservoir geophysics can help in management of fractured reservoirs. If there is a question about local fracture orientation around wells, one can use shear-wave splitting to measure time and amplitude differences; these show fracture orientation. Fractures smaller than the wave scale of seismic data (subseismic fractures) are aligned with those that can be seen in seismic images, but they are considerably more numerous. Where fractures are vertical, they exhibit anisotropy that varies with azimuth. Seismic waves that travel parallel to fractures tend to be faster than those that travel perpendicular to fractures.

Seismic data also can measure the anisotropy caused by fractures from amplitude attributes. The amplitude of the seismic waves is often higher parallel to fractures. Compressional waves (P-waves) can be somewhat sensitive to fractures, but shear waves (S-waves) are especially useful for making that velocity-difference measurement. Generally, they best show the effects of azimuthal anisotropy (Figure 16). Azimuthal anisotropy has been shown to document fracture presence, density, and orientation (Figure 17). That information is very important in the management of oil fields in fractured carbonates because fractures can comprise the majority of the hydrocarbon delivery system.

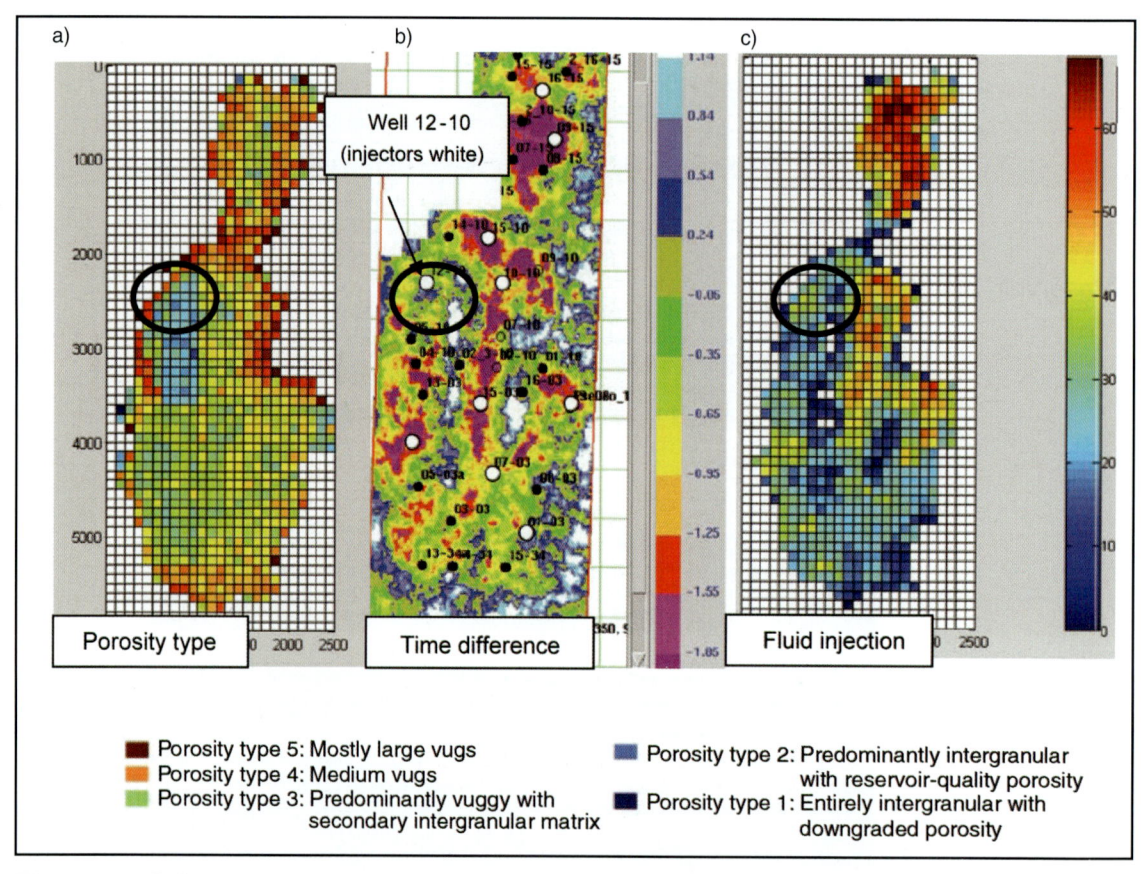

Figure 15. Seismic monitoring of gas and solvent injection into a carbonate reservoir. Porosity type affects the ability to monitor the injection. Low-aspect-ratio pore types (3, 4, and 5 as mapped in [a]) give a good seismic response to the injection. Note (b) the correlation of the seismic time difference from monitoring with (c) the engineering map of the injected fluid. However, where the pore type has low-aspect ratio (type 1 and 2 in [a]), the time difference is suppressed significantly, as at injector location 12-10 (Ng et al., 2005).

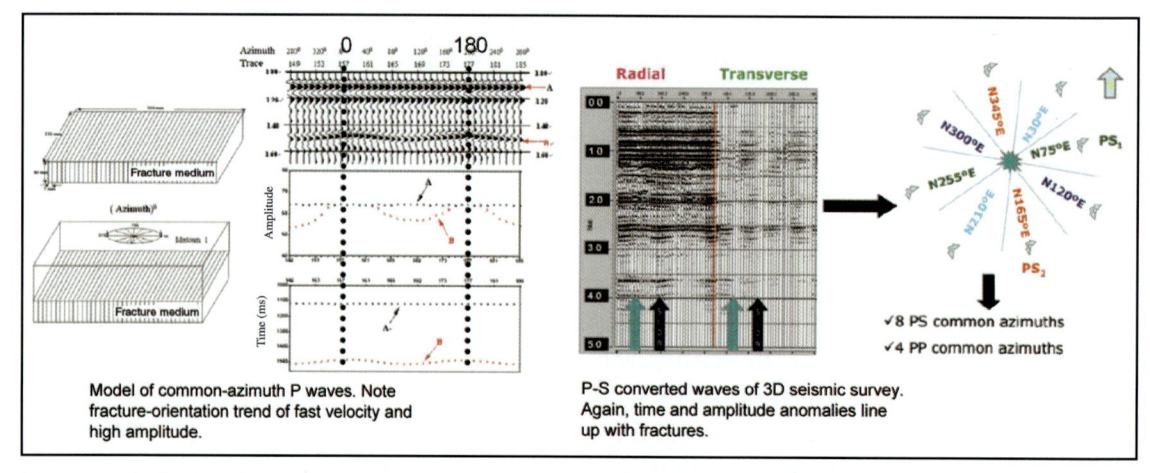

Figure 16. P-waves are faster parallel to fractures. If seismic data are acquired over a full azimuth range, fracture orientations can be interpreted. Some seismic waves that start as P-waves convert a significant part of the reflected energy into shear waves (S-waves), which move from side to side. S-waves are also faster along the orientation of fractures (Guan et al., 2006; Vetri et al., 2003).

## Example of reservoir monitoring in fractured carbonate rocks

Mississippian carbonate rocks of the Weyburn field in western Canada have been through primary production and secondary water flood from vertical wells, but oil was bypassed in the upper, marly carbonate section where porosity is low (Li, 2003). Now the field is being injected simultaneously with gas and solvents to produce bypassed oil in this fractured upper member. The tertiary injectors (horizontal producers and horizontal gas injectors) are aligned along the known fracture trend, which produces a "line-drive" injection effort toward the horizontal producers (Figure 18).

Compared with the base seismic survey when no injection has been activated, repeat 3D seismic surveys show that gas and solvent cause traveltime-delay anomalies below the stratigraphic level where gas enters and flows into the fractured carbonate reservoir (Figure 19). Therefore, to see seismic anomalies lined up northeastward along both the horizontal injection and the fracture trends (Figure 20) is not surprising. The horizon

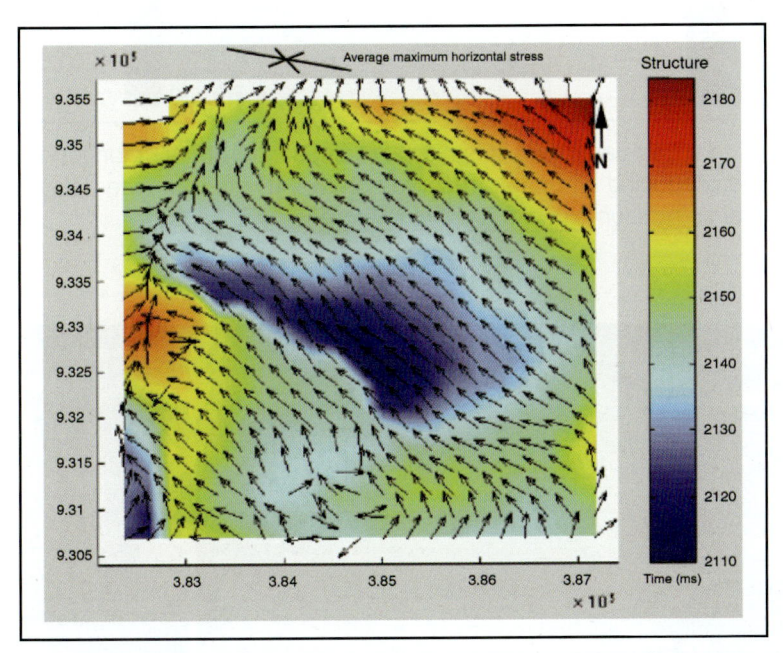

Figure 17. Orientations of fractures in carbonate rocks in the Barinas Apure, Venezuela, from P-S mode conversion. The data are reported to show that fracture orientations are consistent with well control (not shown) (Perez et al., 1999).

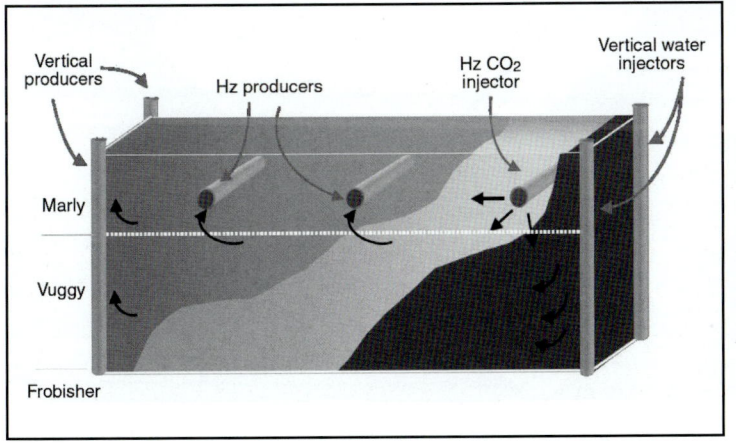

Figure 18. A method of secondary recovery by vertical and horizontal wells was used in carbonate rocks of the Weyburn field, Canada. Oil in the upper marly carbonate facies mostly was bypassed during production by vertical wells. The marly facies is now producing from horizontal wells with horizontal carbon-dioxide injectors (simultaneous but separate water and gas injection [SSWG]) (Li, 2003).

slice of the injection horizon (Figure 21) also shows that amplitude anomalies are aligned northeastward. However, an anomalous amplitude pattern also is aligned northwestward. That pattern shows evidence of movement of injection gas and solvent through orthogonal fractures. That is not the desired drainage pattern for the field.

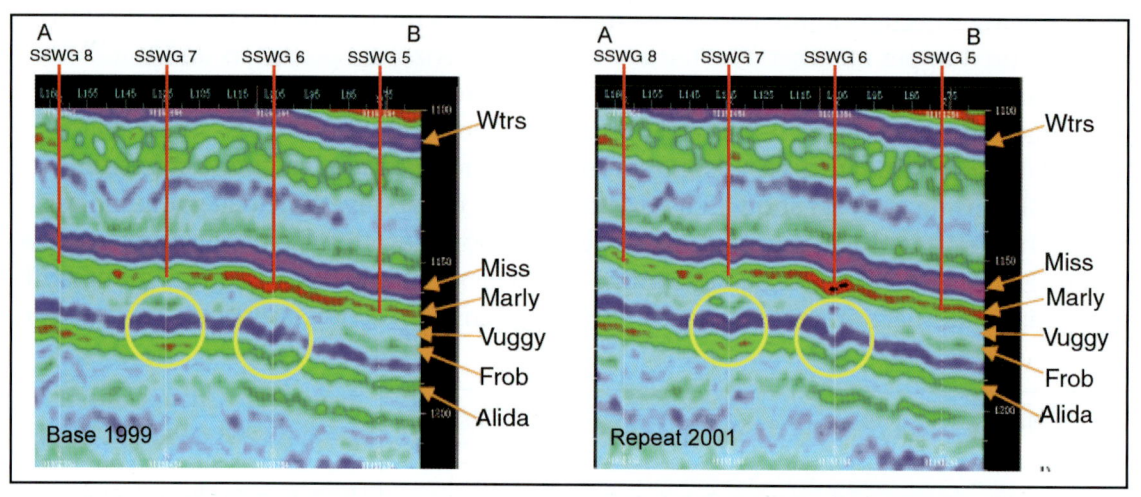

Figure 19. Weyburn field, Canada. In a comparison of seismic cross sections from 4D monitoring, gas injection in the marly part of the producing carbonate reservoir shows 4D seismic time delays from horizons below. The marly seismic horizon exhibits amplitude increase (Li, 2003).

Figure 20. Weyburn field, Canada. The time-delay map of the interval below the marly section shows the spatial distribution of gas injection (Li, 2003).

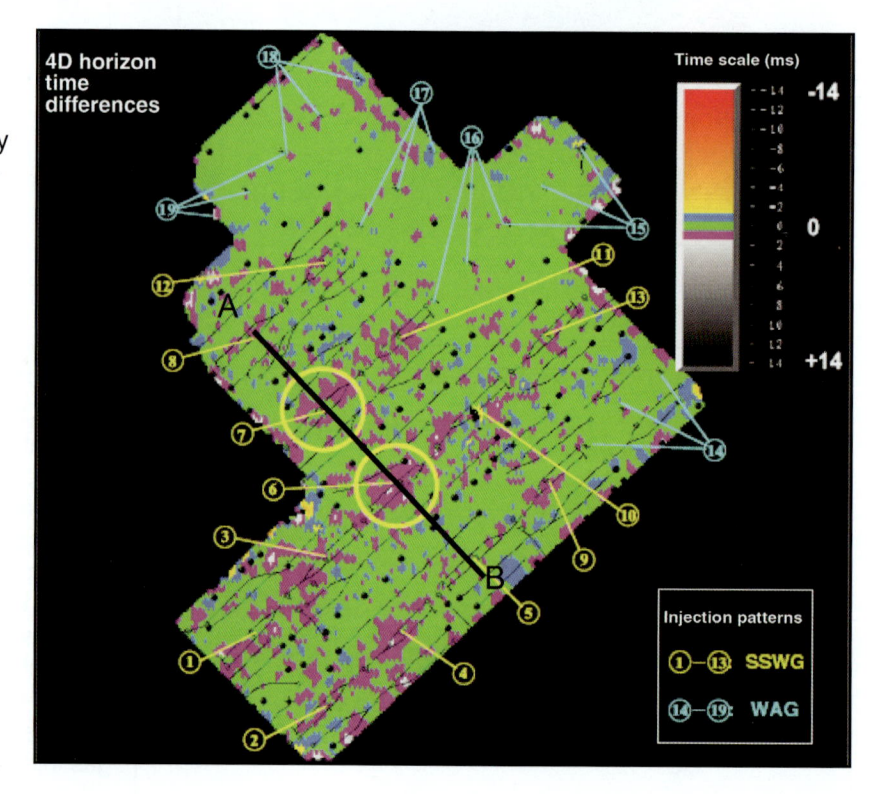

The results of 4D seismic described above are consistent with reservoir-monitoring information taken from well data. Moreover, seismic-amplitude monitoring is confirmed to be a valid method in two additional ways. First, the overlying anhydrite seal shows no consistent amplitude trend, which confirms that the seal is retaining the sequestered $CO_2$. Second, amplitude anomalies observed in the reservoir were confirmed by observed traveltime delays, as predicted from rock physics. Because of the historical pessimism about use of seismic data in monitoring carbonate-rock reservoirs, the examples from the Weyburn field are especially encouraging.

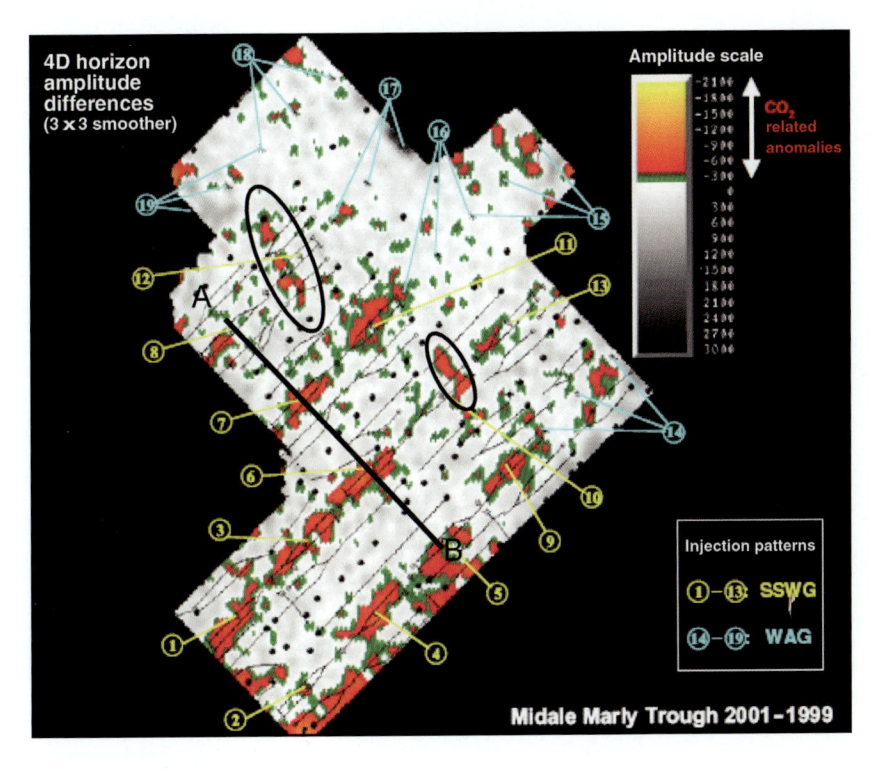

Figure 21. Weyburn field, Canada. The horizon slice of amplitude from the marly reflector shows anomalies that trend along the orientation of horizontal production wells and the primary fracture direction. However, a secondary orthogonal fracture direction also might be operative. That would influence the direction of movement of gas and solvent and the sweep of the reservoir (Li, 2003).

# Appendix A  Salinas Problem

## Delineation results

Now compare our thinking with the actual delineation program in the case described. Three delineation wells were drilled in the year after discovery. Two were dry holes (102 #1 and 103 #2), and one (103 #2 sidetrack [s/t]) was successful in drilling the gas-water contact. The three wells are posted on the horizon slice (Figure A-1). All are south of the 103 #1 discovery well.

The first delineation well (103 #2) was drilled outside the terrane of salt to test the limits of the gas downdip. The well is spotted in a high-amplitude area, less than 100 ft below the proven base of gas in the discovery well. This delineation well penetrated 70 ft of sand but no gas (Figure A-2). From that result, the conclusion was that the gas did not extend far to the south and that moderate seismic amplitude expression alone was not sufficient to determine gas reservoir.

The second delineation well, drilled as a sidetrack to the northeast (103 #2 s/t), penetrated 71 ft of sand, including 12 ft of the gas column above the gas-water contact at 11,343 ft (Figure A-3). This gas-water contact, which is 32 ft lower than the base of the gas in the discovery well, established a critical parameter for the volume of the reservoir. Then it was important to determine the western extent of the reservoir.

The third delineation well (102 #1) was drilled to the west to test the hypothesis of reservoir and gas in the adjacent lease. The 102 #1 delineation well, drilled through salt, was west and south of the discovery well. The delineation well drilled essentially no sand

**Figure A-1.** Poststack-migrated amplitude map showing locations of the three delineation wells drilled (103 #2, 103 #2 sidetrack [s/t], and 102 #1). Note that only the 103 #2 sidetrack contained any hydrocarbons. The other wells were dry holes, apparently positioned outside the reservoir.

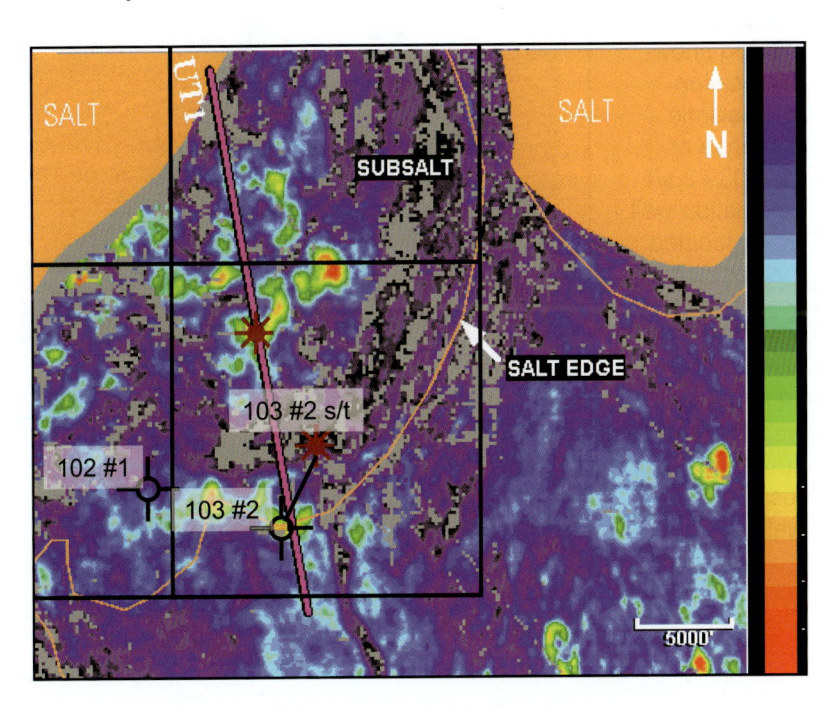

at that location and showed that the reservoir has a stratigraphic limit between that location and the other wells (Figure A-1).

## Development plan and seismic upgrade

Plans were made then for development of the field by completing the discovery well (103 #1) as a producer and by adding a few nearby development wells to exploit

**Figure A-2.** Seismic line and log showing location and seismic response of the 103 #2 dry delineation well drilled outside salt. Although the sand is well developed, no hydrocarbons are present.

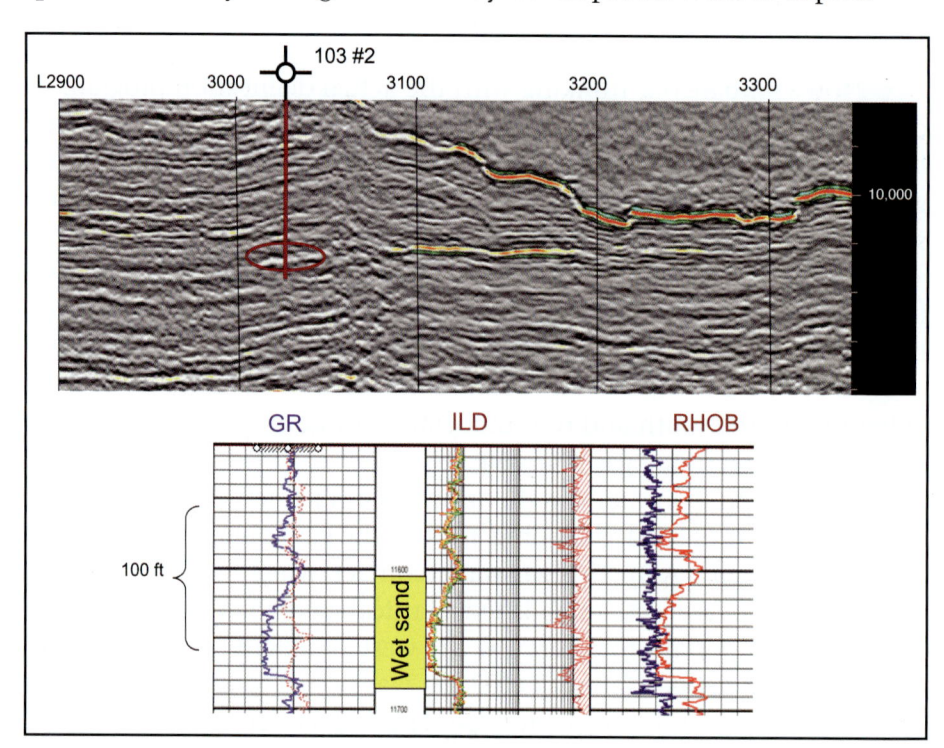

**Figure A-3.** Seismic line and log showing location and seismic response of the 103 #2 s/t delineation well, which located the gas-water contact and a very short column of gas. Note the very weak but measurable seismic response at the well.

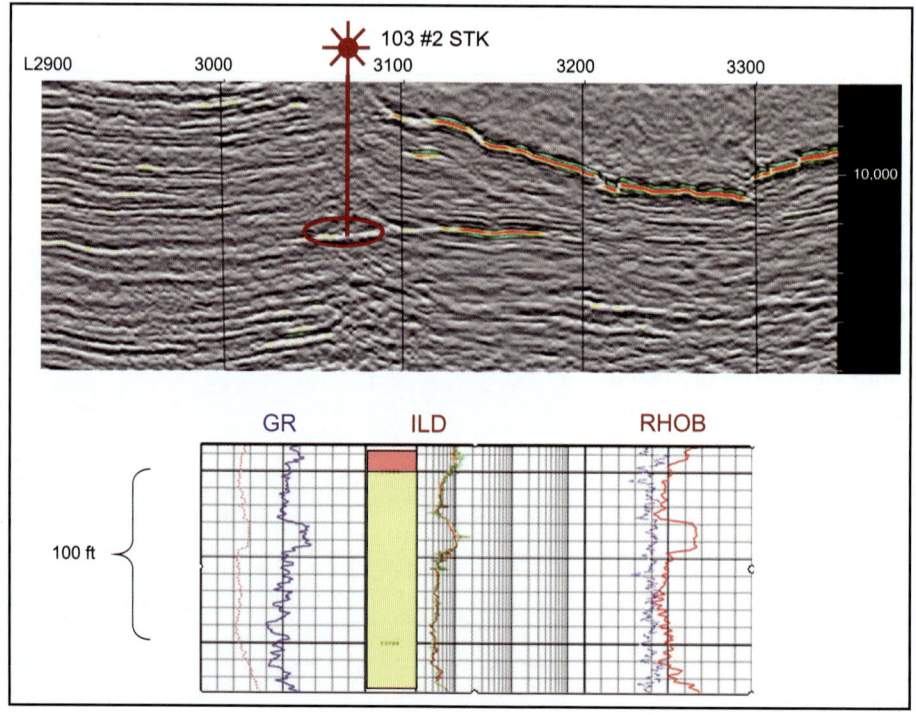

compartments of the reservoir. The seismic data certainly suggest that the reservoir could be separated east and west by a seal within the central channel. The channel might have been an abandonment channel filled with shale. To spot the development wells correctly, it was also important to understand the stratigraphic variations northward and southward.

During the time required to prepare for development wells and install production facilities, the geophysics team significantly improved the velocity field to image the seismic data and then performed a prestack depth migration. The result of the horizon slice compared with the poststack data (Figure A-4) shows the marked improvement of seismic imaging. The new image suggests a stratigraphic model of a restricted sand fairway with a centralized channel.

If the amplitude information is even moderately close to being correct, it shows that the reservoir is narrow, with strong likelihood of compartmentalization. The impact of prestack depth imaging was successful. Noise from subsalt multiples was handled, and prestack data were positioned correctly before stacking. Comparison of prestack seismic trace gathers outside salt and under salt (Figure A-5) shows serious multiple noise. A good demultiple program was required for removal of the multiples. Comparison of cross sections of poststack and prestack data shows that the prestack 3D imaging is a significant improvement (Figure A-6). When delineation wells are posted on the 3D prestack depth-migrated horizon slice (Figure A-7), it becomes clear that the amplitude attribute from seismic data is a good indicator of where the reservoir is present.

## Development

Based on the improved seismic data, development wells #3 and #4 were drilled. They penetrated 76 and 109 ft of reservoir pay, respectively, in different compartments, as confirmed by different pressures (Figures A-8 through A-10). The three development

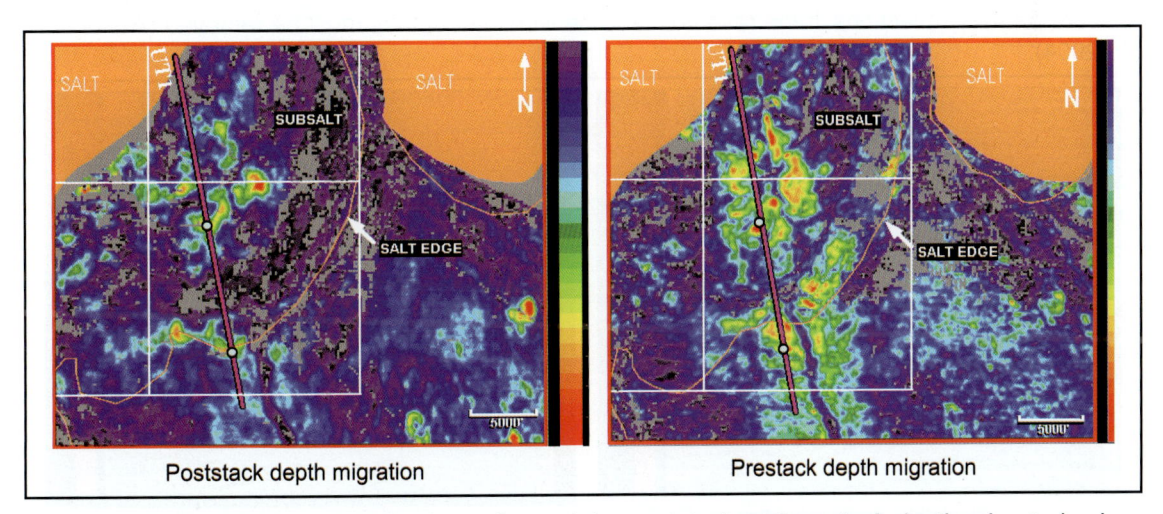

Figure A-4. Comparison of the horizon slices of the poststack and prestack depth-migrated seismic data shows the improvement of seismic imaging. The reservoir appears to be confined to a channel about 10,000 ft wide beneath and outside salt. Note that the horizon also is illuminated better outside the terrane of salt (Abriel et al., 2004).

wells (#1, #3, and #4) were put on stream for production, and the field began to generate revenue. The field produced approximately as predicted. During the first seven-year production interval, income eventually covered the cost of exploration, development, and production, and the field made an acceptable profit (Figure A-11).

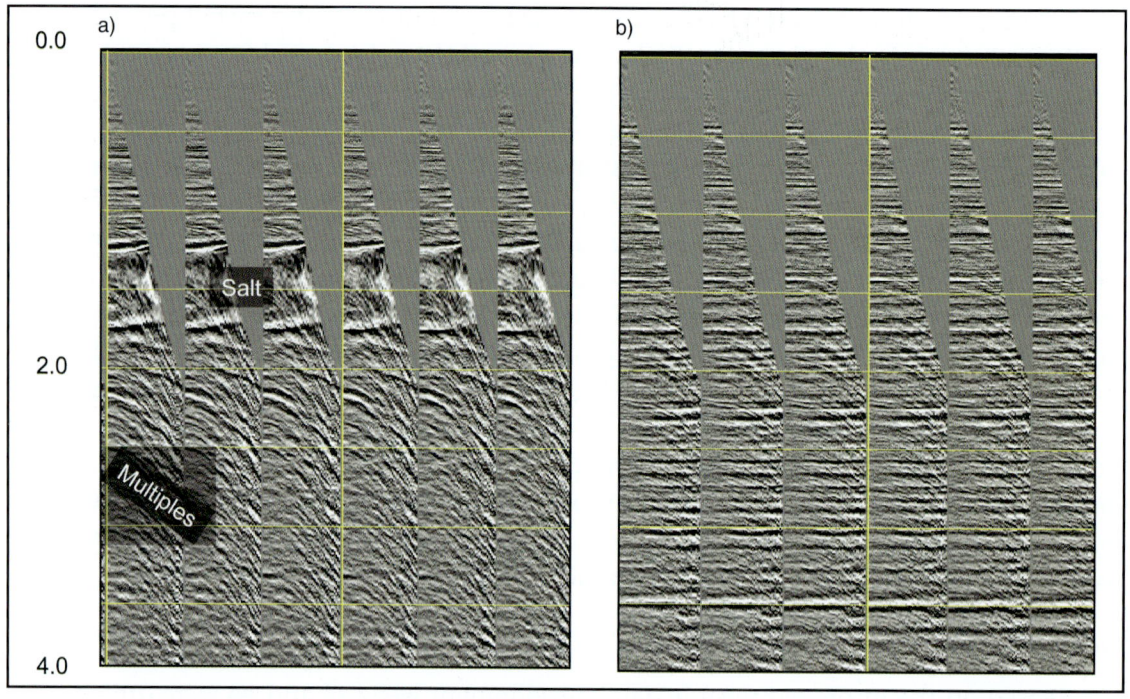

Figure A-5. Comparison of prestack depth-migrated data (a) under salt and (b) outside salt shows one of the problems with subsalt data. Prestack gathers should be flat prior to stacking. The parabolic events below salt are multiples. They are so strong that they almost completely obscure primary reflections. If the reservoir is to be illuminated, they must be removed (Abriel et al., 2004).

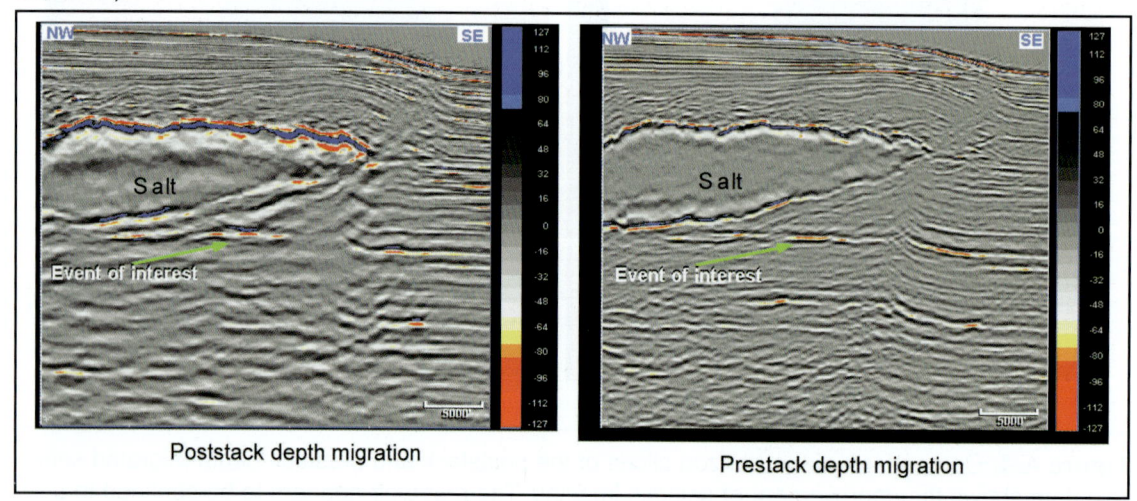

Figure A-6. Comparison of seismic sections of prestack and poststack depth-migrated data shows the reservoir to have been imaged much better. The reservoir is of coherent shape and size, and the entire subsalt image is much improved (Abriel et al., 2004).

## Reservoir model

A reservoir flow-simulation model was constructed prior to development drilling and was updated prior to starting production. The reservoir response was checked periodically to be sure that compartmentalization was estimated correctly and that water could be managed as anticipated. The simulator was built with knowledge of the reservoir from well data and from seismic data. To depict the geology better for the reservoir model,

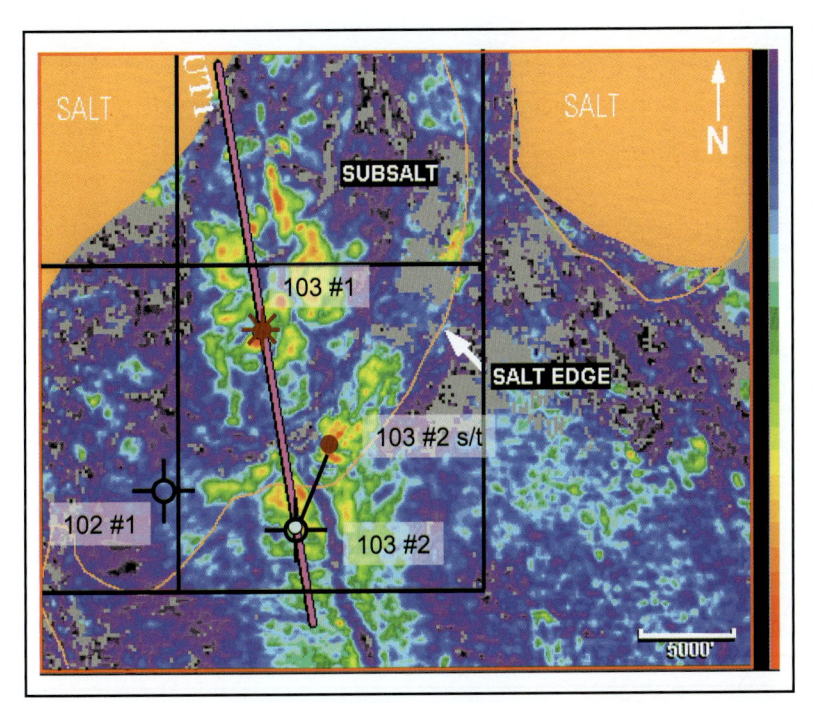

**Figure A-7.** The three delineation wells are shown on the prestack-migrated horizon slice. The 103 #1 discovery well is under salt on the western side of the channel complex. The 102 #1 dry hole is outside the channel complex. Dry hole 103 #2 is in the channel but is wet, and the 103 #2 sidetrack is just at the gas/water contact.

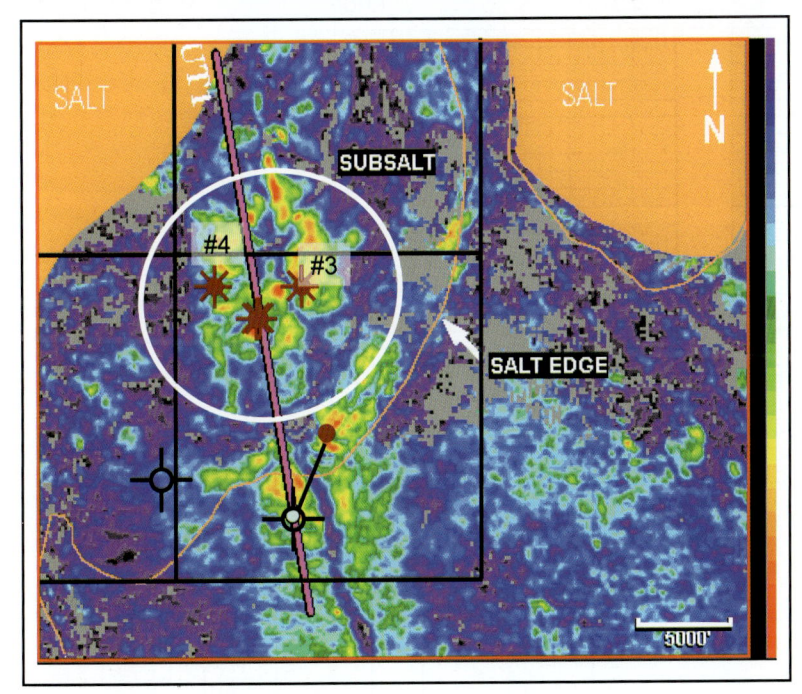

**Figure A-8.** The three development wells used for producing the reservoir are shown on the prestack-migrated horizon slice. The wells were designed to drain compartments, which were indicated by seismic amplitude.

seismic data were modified by an "inversion" process that attempts to use well data and back away the wave nature of the seismic data. Thus, the data more closely resemble impedance logs.

**Figure A-9.** Prestack depth-migrated cross sections show locations of new development wells and amplitude continuity of the reservoir.

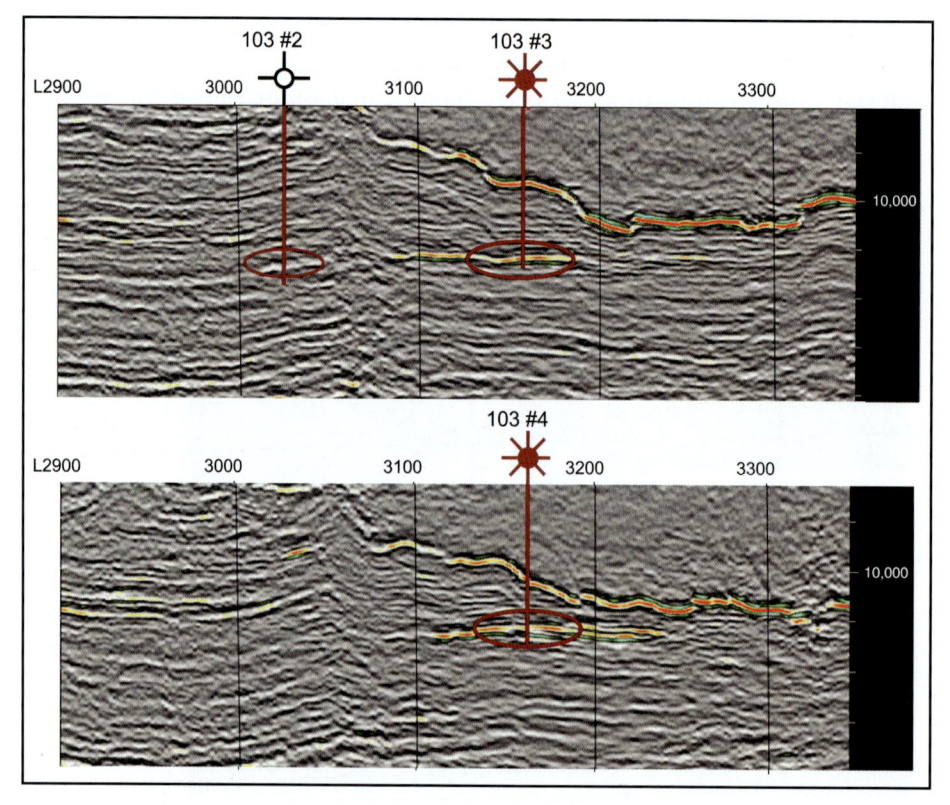

**Figure A-10.** Logs of additional development wells show thick reservoir, full of gas.

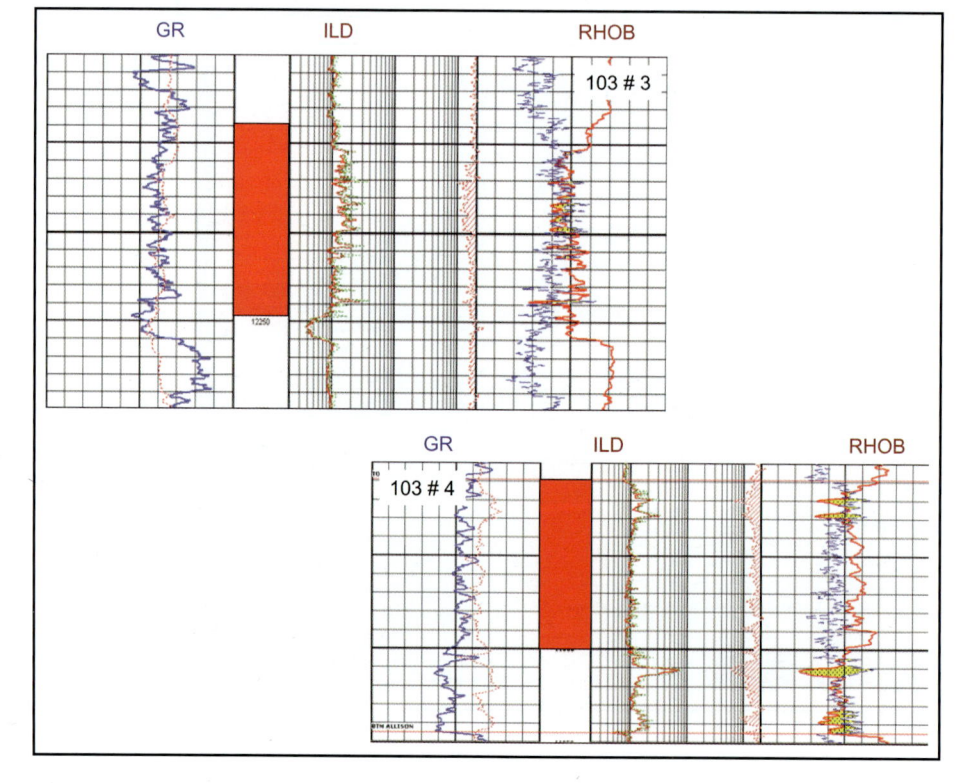

A cross section through the simulator model (Figure A-12) shows the influence of the well information, the geologic conceptual model, and the inverted seismic data used to describe reservoir heterogeneity.

## Remarks

Although this project is not exceptionally large, it serves a very useful purpose in illustration of the use of geophysics in delineation. As described, seismic data were required to image the reservoir where it is beneath the salt. During development of the field, continuous work on the seismic data provided images that influenced the discovery, delineation, and development well locations as well as the description of reservoir structure, compartmentalization, net pay, porosity, and boundaries of the fluids. In comparison with the cost of drilling wells, the cost of seismic data is low. Seismic data provided information about the subsurface that was critical to the success of the project. Without judicious use and continuous improvement of seismic data, it would not have been possible to make an economic field from this reservoir.

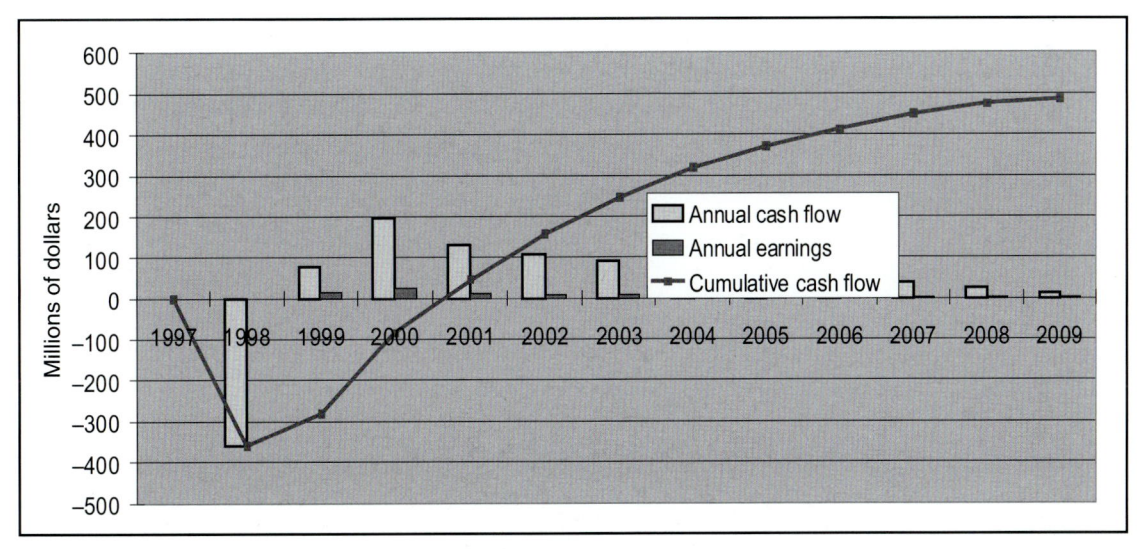

Figure A-11. Cash flow of this project from development through production shows a high initial expense followed by a return on investment sufficient to make the project profitable.

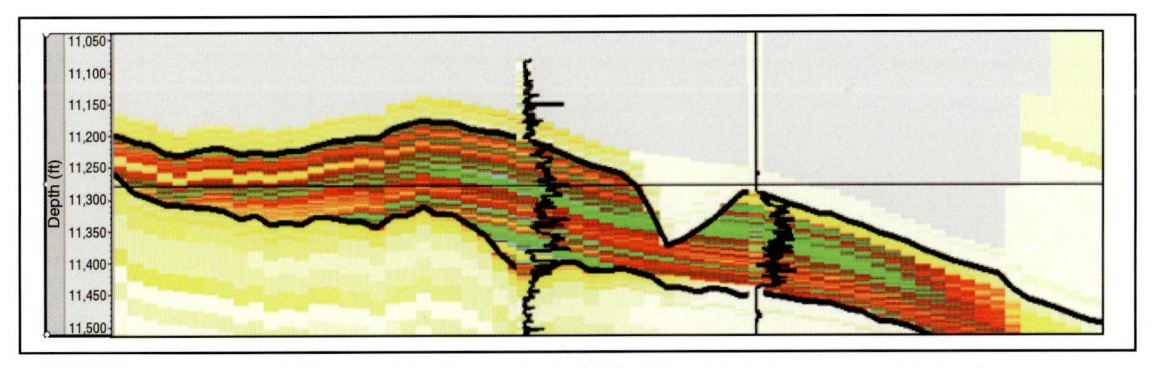

Figure A-12. A cross section of the reservoir model shows porosity (maximum shown in green). The engineering flow model was influenced strongly by seismic response. Seismic data and well data were used simultaneously to constrain reservoir parameters.

# Appendix B   Cobra Problem

## Results

Compare your reasoning and conclusions with the development program that went forward. Three development wells were drilled in years 24 through 27. Wells #19, #20, and #21 (Table B-1) were drilled updip from the water injectors and were put into production (Figure B-1). All three penetrated the expected oil leg. As spotted on the seismic-amplitude map (Figure B-2), those wells are safely in high-amplitude areas (red). The prediction was that they should have high oil saturations, should be far updip of the average water level, and should be pressure-supported in part by the injector well.

Table B-1. Development wells drilled after the first 3D survey.

| Well | Date | Gas (ft) | Oil (ft) | Water (ft) |
|------|---------|----------|----------|------------|
| 19 | Year 24 | 30 | 82 | 0 |
| 20 | Year 24 | 10 | 84 | 14 |
| 21 | Year 27 | 0 | 103 | 0 |

### Second 3D seismic survey in year 27

After nine years of using 3D seismic data to help manage the reservoir in this fault block and in the rest of the field, a second 3D survey was acquired. The second survey was designed to improve the overall seismic image of the reservoir and possibly to monitor the time-lapse (4D) seismic effects of production (Figure B-3). Time-lapse effects that were considered included water encroachment, evolution of water-injection pathways, gas coming out of solution during pressure drop, and bypassing of oil-bearing compartments.

Several factors contributed to the second survey's not providing an ideal 4D seismic view of the reservoir. First, the second survey did not use the acquisition parameters of

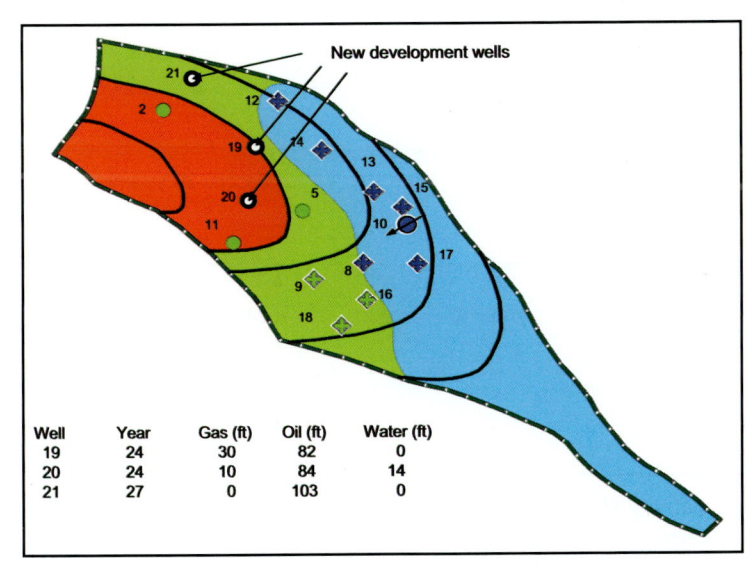

**Figure B-1.** The reservoir was developed with three additional wells, represented by white dots. All three were put into production successfully.

the first survey because 4D was not the primary purpose of reacquisition. Further, the second survey was affected even more strongly by noise and difficulty of access, partly because of increased activity in production of the field during acquisition of seismic data. The seismic data from year 18 and the data from year 27 were processed poststack. Processing the data sets together prestack would have been preferred, but the option was not available. Those issues are unfortunate but not unique. Live projects are like that.

Despite the seismic data sets not having been optimized for 4D analysis, extraction of time-lapse attributes for understanding the reservoir was attempted. Data from both surveys were processed poststack using time, amplitude, and phase shifts of the seismic traces (cross-equalization). The objective was to produce data sets that would show little difference in attributes outside the reservoir.

**Figure B-2.** The amplitude map of seismic data from year 18 shows how the three new development wells were located in high-amplitude areas and about 1 km from other producers. To avoid the gas cap, none was placed in the west.

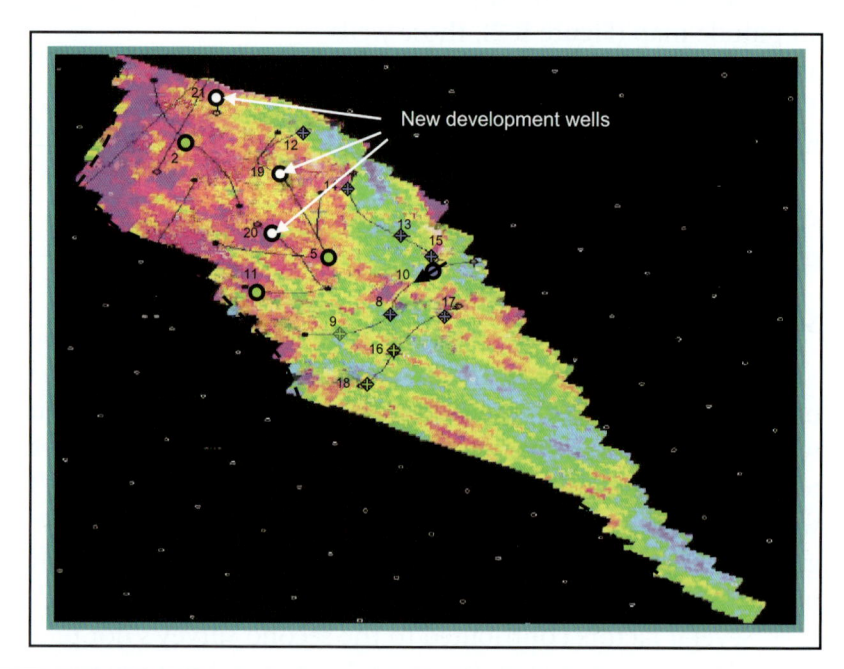

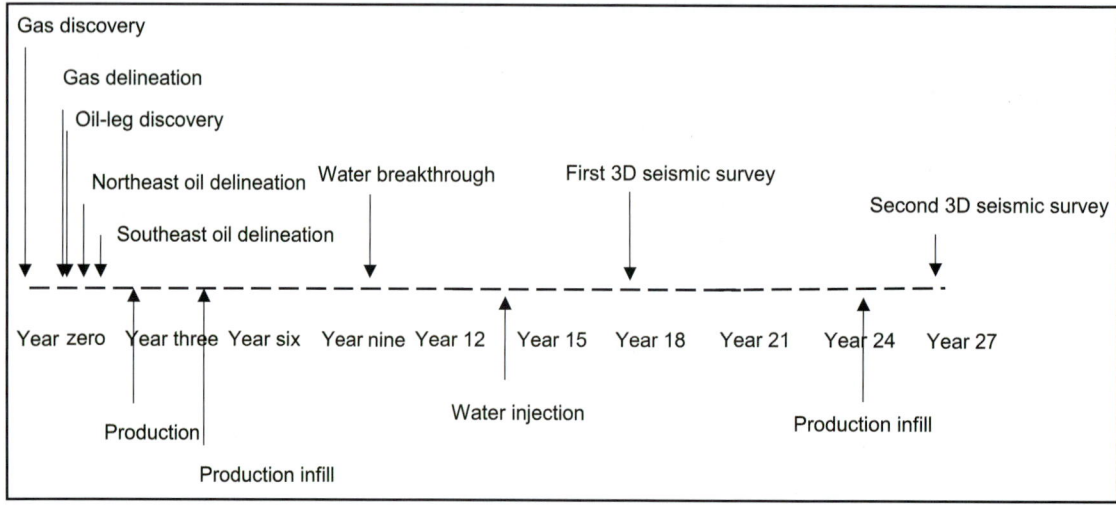

**Figure B-3.** The time line of reservoir activity shows the production infill several years after the first 3D survey and acquisition of a 3D survey in year 27.

Side-by-side comparison of those processed surveys appears to show the effect of nine years of water injection and oil production (Figure B-4). Areas least affected by production are downdip of water injectors and at the northwestern corner. Areas of greatest seismic difference are those where water apparently has moved updip in connected fingers along the northeastern fault and in the south-central part of the fault block.

## Interpretation

Interpretation of the reasons for uneven sweep of the reservoir (Figure B-5) includes faulting and variation in stratigraphy. Based on previous structural analysis, intrareser-

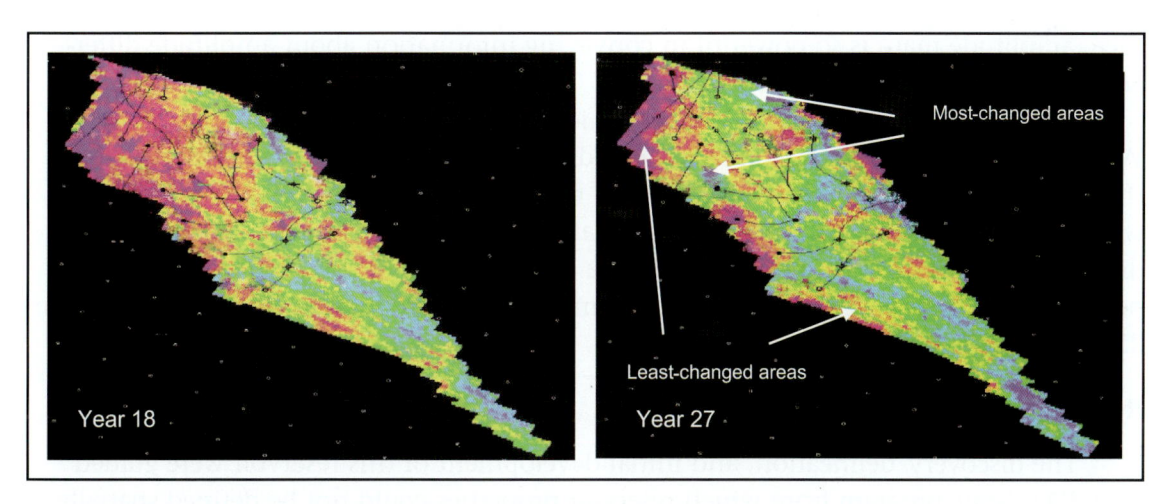

Figure B-4. Comparison of the horizon slices from the separate 3D surveys of year 18 and year 27 shows the extension of low amplitudes within the producing area. The suggestion is strong that this extended pattern of low amplitude is evidence of the advance of water.

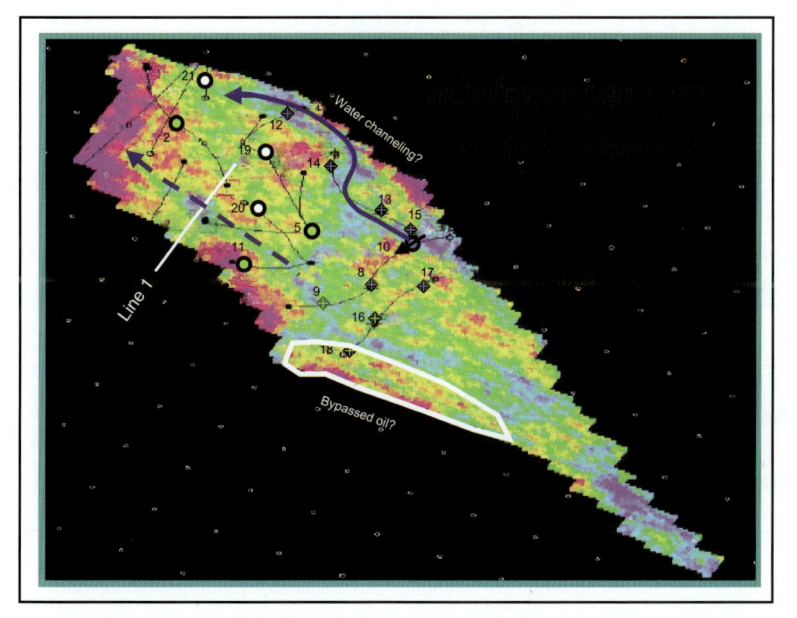

Figure B-5. The amplitude map from the survey of year 27 shows locations of the six remaining producers (green and white dots) and of the water injector. The most probable interpretation is that water has moved into the producing area along stratigraphically connected flow channels in the low-amplitude zones. Part of the reason for bypass also might be impedance of flow by small faults in the reservoir. The location of line 1 is referenced for Figure B-6.

voir faults trend northwestward. Faults seem to be barriers to flow of oil and water and thereby explain why oil might have been bypassed downdip, as at well #18. Updip, water apparently is replacing oil and flowing in preferential "channels." These are the connected sand stratigraphy of a deltaic depositional system. The six remaining oil producers are not located directly in those channels and so are producing mostly oil in the remaining higher-amplitude areas.

A cross-sectional view (Figure B-6) of the seismic data of the first and second surveys is referenced from the amplitude map (Figure B-5). This cross section shows significant decrease of amplitude in the reservoir zone. When seismic data are viewed only in cross-sectional form (lines), those effects are difficult to understand. That is why the horizon slice (amplitude map) is so powerful in conveying information about amplitude differences in seismic data.

Another representation of the 4D seismic data is a difference volume (Figure B-7). When the technology works correctly, large differences should be detectable only in the area of the produced reservoir. Although the image is imperfect, that is generally what we see in this project. The difference data also can be sliced along the horizon and displayed. The horizon slice of the difference cube (Figure B-8) shows information that permits strong inferences about water movement during production from year 18 to year 27.

## Remarks

The discovery, delineation, and initial development of this reservoir were guided by a 2D seismic program from which reservoir properties could not be defined spatially. Once production was under way and 3D seismic data were acquired, geology, geophysics, and engineering data were combined in a team effort to manage the reservoir. Faults and differences in facies seem to account for uneven sweep of the reservoir during primary and secondary recovery. Both 3D and 4D seismic data assisted in explaining production histories of wells, providing opportunity to manage production better.

**Figure B-6.** A seismic line from the 3D surveys shows decreased amplitude in the reservoir zone, interpreted to be evidence of water channeling.

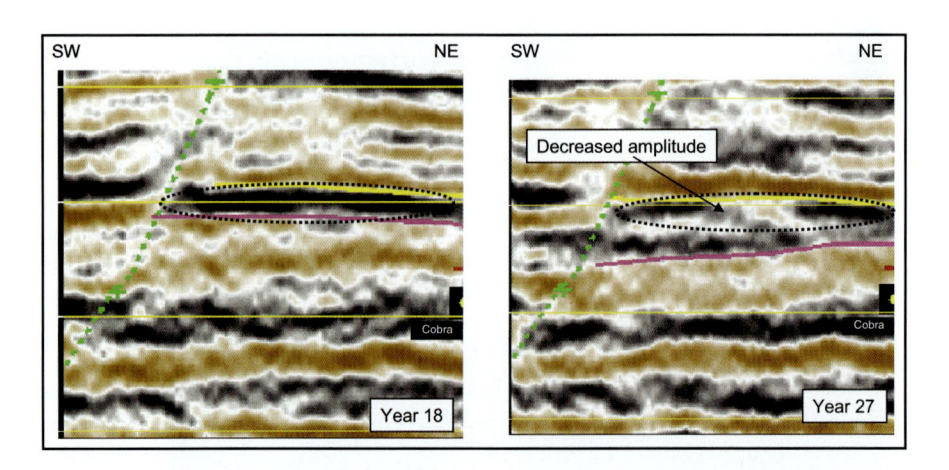

*Appendix B   Cobra Problem*

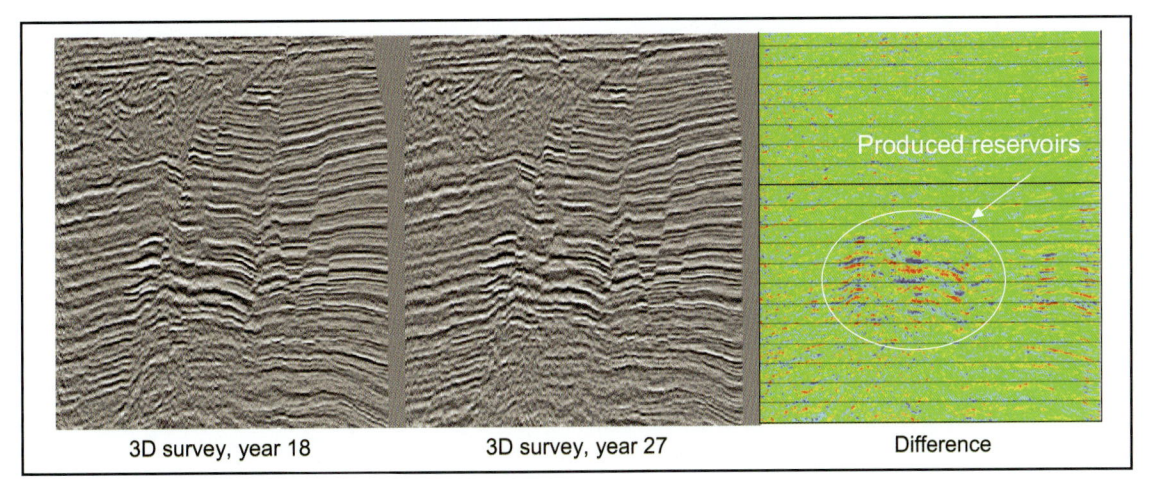

**Figure B-7.** The 3D surveys from year 18 and year 27 were cross-equalized poststack. Although the surveys were not designed to be used primarily in time-lapse analysis, the difference of the two surveys shows evidence of changes in several produced reservoir zones.

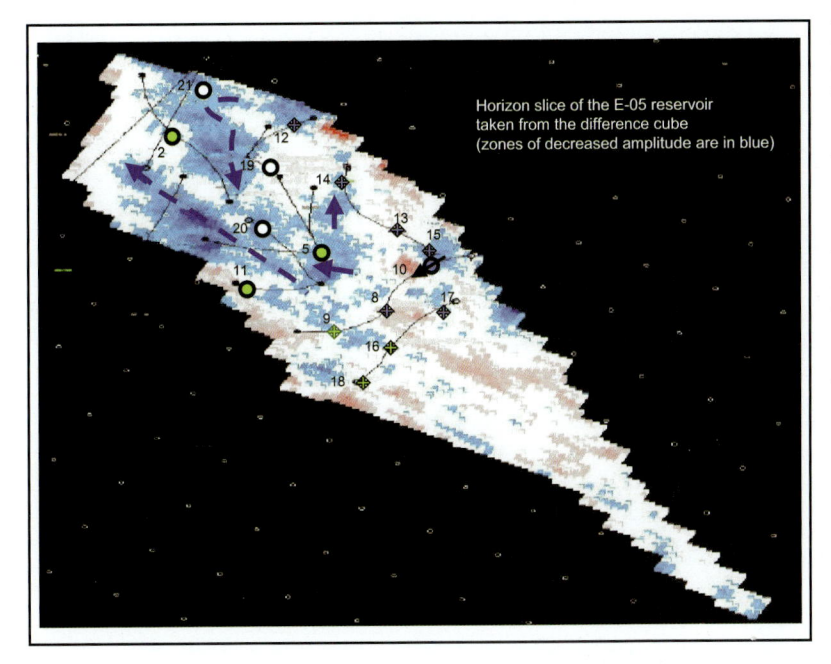

**Figure B-8.** The horizon slice of the difference of the two surveys shows significant decrease in amplitude in the updip, produced part of the reservoir. The most reasonable interpretation is that water replaces oil along flow channels. The uneven configuration of the differences could be the effect of stratigraphic heterogeneity.

*Distinguished Instructor Short Course* • 119

# References

Abriel, W. L., and R. M. Wright, 2004, Seismic data interpretation for reservoir boundaries, parameters, and characterization, *in* A. R. Brown, Interpretation of three-dimensional seismic data, sixth edition: AAPG Memoir 42 and SEG Investigations in Geophysics Series No. 9, 366–374.

Abriel, W., J. Stephani, R. Shank, and D. Bartel, 2004, 3-D depth image interpretation, *in* A. R. Brown, Interpretation of three-dimensional seismic data, sixth edition: AAPG Memoir 42 and SEG Investigations in Geophysics Series No. 9, 449–475.

Abriel, W., P. Neale, J. Tissue, and R. Wright, 1991, Modern technology in an old area: Bay Marchand revisited: The Leading Edge, **10**, 21–35.

Bassant, P., F. Van Buchem, A. Strasser, and A. Lomando, 2004, A comparison of two early Miocene carbonate margins: The Zhujiang carbonate platform (subsurface, South China Sea), and the Pirinc platform (outcrop, southern Turkey), *in* G. M. Grammer, P. M. Harris, and G. P. Eberli, eds., Integration of outcrop and modern analogs in reservoir modeling: AAPG Memoir 80, 153–170.

Belopolsky, A., and A. Droxler, 2003, Imaging Tertiary carbonate system — The Maldives, Indian Ocean: Insights into carbonate interpretation: The Leading Edge, **22**, 646–652.

Bergeon, T., W. Abriel, J. Rafalowski, and B. Regal, 1993, Green Canyon block 205: Geophysical analysis of a deepwater Gulf of Mexico discovery: 63rd Annual International Meeting, SEG, Expanded Abstracts, 437–438.

Clark, M., L. Klonsky, and K. Tucker, 2001, Geologic study and multiple 3-D surveys give clues to complex reservoir architecture of giant Coalinga oil field, San Joaquin Valley, California: The Leading Edge, **20**, 744–751.

Curtis, C., R. Kooper, E. Decoster, A. Guzman-García, C. Huggins, A. Knauer, M. Minner, N. Kupsch, L. Linares, H. Rough, and M. Waite, 2002, Heavy oil reservoirs: Oilfield Review, 30–51.

Dasgupta, S., M. Hong, and I. Al-Jalal, 2001, Reservoir characterization of Permian Kuff-C carbonate in the supergiant Ghawar field of Saudi Arabia: The Leading Edge, **20**, 706–717.

Droste, H., and M. Van Steenwinkle, 2004, Stratal geometries and patterns of platform carbonates: The Cretaceous of Oman, *in* G. Eberli, J. Masaferro, and J. Sarg, eds., Seismic imaging of carbonate reservoirs and systems: AAPG Memoir 81, 185–206.

Dubucq, D., F. Lefeuvre, and F. Bertini, 2003, Deep offshore seismic monitoring: The Girassol field, a West Africa textbook example: 73rd Annual International Meeting, SEG, Expanded Abstracts, 1338–1341.

Eberli, G, G. Baechle, F. Anselmetti, and M. Incze, 2003, Factors controlling elastic properties in carbonate sediments and rocks: The Leading Edge, **22**, 654–660.

Fischer, K., M. Holling, R. Marchall, and J. Mau, 1997, Remarks on exploration tools: Integrated exploration strategy applied to carbonate environments, in L. Thomsen, I. Palaz, and K. Marfurt, eds., Carbonate seismology: SEG Geophysical Developments Series No. 6, 203–222.

Fischer, K., U. Moller, and R. Marchall, 1997, Development of an exploration concept for the Schuaiba formation using seismic sequence and facies analysis with forward modeling, in L. Thomsen, I. Palaz, and K. Marfurt, eds., Carbonate seismology: SEG Geophysical Developments Series No. 6, 407–416.

Fournier, F., and J. Borgomano, 2007, Geological significance of seismic reflections and imaging of the reservoir architecture in the Malampaya gas field (Philippines): AAPG Bulletin, **91**, 235–258.

Gratwick, D., and C. Finn, 2005, What's important in making far-stack well-to-seismic ties in West Africa?: The Leading Edge, **24**, 739–745.

Guan, L., S. Wang, and H. Zsu, 2006, Prediction of a fracture-cavern system in a carbonate reservoir: A case study from Tahe oil field, China: The Leading Edge, **25**, 1396–1400.

Handford, C., 1998, Visualizing exploration targets through carbonate sequence stratigraphy: The Leading Edge, **17**, 891–894.

Hoversten, M., P. Milligan, J. Byun, J. Washburne, L. Knauer, and P. Harness, 2004, Crosswell electromagnetic and seismic imaging: An examination of coincident surveys at a steam flood project: Geophysics, **69**, 406–414.

Jenkins, S., M. Waite, and M. Bee, 1997, Time-lapse monitoring of the Duri steamflood: A pilot and case study: The Leading Edge, **16**, 1267–1273.

Kaderali, A., M. Jones, and J. Howlett, 2007, White Rose seismic with well data constraints: A case history: The Leading Edge, **26**, 742–754.

Larue, D. K., and Y. Yue, 2003, How stratigraphy influences oil recovery: A comparative reservoir database study concentrating on deepwater reservoirs: The Leading Edge, **22**, 332–339.

Latimer, R., R. Davidson, and P. van Riel, 2000, An interpreter's guide to understanding and working with seismic-derived acoustic impedance data: The Leading Edge, **19**, 242–256.

Li, G., 2003, 4D seismic monitoring of $CO_2$ flood in a thin fractured carbonate reservoir: The Leading Edge, **22**, 690–695.

Li, Y., J. Downton, and B. Goodway, 2003, Recent applications of AVO to carbonate reservoirs in the Western Canadian Sedimentary Basin: The Leading Edge, **22**, 670–674.

Linari, V., M. Santiago, C. Pastore, K. Azbel, and M. Poupon, 2003, Seismic facies analysis based on 3D multiattribute volume classification, La Palma field, Maracaibo, Venezuela: The Leading Edge, **22**, 32–35.

Lindseth, R., 1979, Synthetic sonic logs — A process for stratigraphic interpretation: Geophysics, **44**, 3–26.

Liu, Y., A. Harding, W. Abriel, and S. Strebelle, 2004, Multi-point simulation integrating wells, 3-D seismic data and geology: AAPG Bulletin, **88**, 905–921.

Lynch, S., and L. Lines, 2004, Combined attribute displays: 74th Annual International Meeting, SEG, Expanded Abstracts, 1953–1956.

Marion, D., A. Nur, H. Yin, and D. Han, 1992, Compressional velocity and porosity in sand-shale mixtures: Geophysics, **57**, 554–563.

Ng, H., L. Bentley, and E. Krebes, 2005, Monitoring fluid injection in a carbonate pool using time-lapse analysis: Rainbow Lake case study: The Leading Edge, **24**, 530–534.

Partyka, G., J. Gridley, and J. Lopez, 1999, Interpretational applications of spectral decomposition in reservoir characterization: The Leading Edge, **18**, 353–360.

Paulsson, B., M. Karrenbach, P. Milligan, A. Goertz, and A. Harding, 2004, High resolution 3D seismic imaging using 3-C data from large downhole seismic arrays: First Break, **23**, 73–83.

Perez, M., V. Grechka, and R. Michelena, 1999, Fracture detection in a carbonate reservoir using a variety of seismic methods: Geophysics, **64**, 1266–1276.

Poupon, M., J. Gil, D. Vannaxay, and B. Cortiula, 2004, Tracking Tertiary delta sands (Urdaneta West, Lake Maracaibo, Venezuela): An integrated seismic facies classification workflow: The Leading Edge, **23**, 909–912.

Powley, D., 2007, Illustrated summary of compartments/pressure regimes in selected North American basins: Part 3: Rocky Mountains, western Canada, and Alaska: AAPG Search and Discovery article #60011, accessed September 28, 2007, http://www.searchanddiscovery.net/documents/2007/07010powley02/images/powley02_07.pdf.

Sheriff, R., 2002, Encyclopedic dictionary of applied geophysics, fourth edition: SEG.

Sigit, R., P. Morse, and K. Kimber, 1999, 4-D seismic that works: A successful large scale application, Duri steamflood, Sumatra, Indonesia: 69th Annual International Meeting, SEG, Expanded Abstracts, 2055–2058.

Skirius, C., S. Nissan, N. Haskell, K. Marfurt, S. Hadley, D. Ternes, K. Michel, I. Reglar, D. D'Amico, F. Deleincourt, T. Romero, R. D'Angelo, and B. Brown, 1999, 3-D seismic attributes applied to carbonates: The Leading Edge, **18**, 384–393.

SPE, 1997, Petroleum reserves definitions, accessed July 14, 2007, http://www.spe.org/spe-site/spe/spe/industry/reserves/Petroleum_Reserves_Definitions_1997.pdf.

Sullivan, M. D., D. Stern, J. L. Foreman, G. N. Jensen, D. C. Jennette, and F. J. Goulding, 2004, An integrated approach to characterization and modeling of deep-water reservoirs, Diana field, western Gulf of Mexico, in G. M. Grammer, P. M. Harris, and G. P. Eberli, eds., Integration of outcrop and modern analogs in reservoir modeling: AAPG Memoir 80, 215–234.

Vetri, L., E. Loinger, J. Gaiser, A. Grandi, and H. Lynn, 2003, 3D/4D Emilio: Azimuth processing and anisotropy analysis in a fractured carbonate reservoir: The Leading Edge, **22**, 675–679.

Tidwell, V., 2007, Sandia National Laboratories, accessed January 11, 2008, http://www.nwer.sandia.gov/wlp/factsheets/light.pdf.

Wilt, M., and M. Morea, 2004, 3D waterflood monitoring at Lost Hills with crosshole EM: The Leading Edge, **23**, 489–493.

# Index

**A**
abandonment channel, 59, 109
acoustic impedance, 47, 50
   for differentiation of seismic response, 50
   transformation of seismic data to, 47
acoustic parameters, earth materials, 43
acoustic response, factors in, 43
acoustic well logs, 47
air guns, 2
algal bindstones/rhodolites, 93
amalgamated channels and systems, 10, 25, 31
amplitude, absolute, and fluid content, 27
amplitude, and fractures, 101
amplitude, correlated with oil and gas saturation, 55
amplitude, effect of gas on, 76
amplitude, effect of oil on, 76
amplitude, for differentiation of seismic response, 50
amplitude anomaly, 29, 31, 32
amplitude attributes, 50, 74, 109
   and prediction of fluids, 74
   and prediction of porosity, 74
amplitude differences, spatial, patterns, 44
amplitude maps, 20, 31, 36, 76, 77, 91, 116, 117, 118
   correlated with reservoir quality, 31
   poststack-migrated, 36
amplitude variation with offset (AVO), 7, 16, 22, 23, 24, 25, 26, 27
   channel deposits, responses, 26
   described, 23
   far-offset amplitudes, 25
   gas caps, detection of, 25
   near-offset amplitudes, 25
   oil, response of, 23
   overbank deposits, responses, 26
   restricted-gradient attribute, 27
      channel-sand facies, 27
      laminated-sand facies, 27
   rock type, response to, 27
   water, response of, 23
amplitude variations and reservoir properties, 7
amplitude with offset, principles, 22
amplitudes and bypassed oil, 60, 61, 62
Angola, 58–59
anhydrite, 2, 105
anisotropy, and fractures, 101
anisotropy, azimuthal, 101, 102, 103
aspect ratio, of pores, 98–100, 102
Athabasca tar sands, 82
attenuation, 16
attribute selection, workflow, 47
AVO (*see* amplitude variation with offset)
AVO attribute, 50, 51
azimuthal anisotropy, 101, 102, 103

**B**
back-reef environments, 92
Barinas Apure, Venezuela, 103
barrier-bar sands, 66
Bay Marchand field, Gulf of Mexico, 60–63
   amplitudes, and bypassed oil, 60, 61, 62
   bypassed oil, 60, 61, 62
   fault blocks, 61, 62
   oil bright spot, 60
   production rates, 63
   reflection response, and water saturation, 60
   reservoir, horizon slice, 61
   swept zones, 60
   3D seismic surveys, dates, 60, 61, 62
      effects of, 61, 62
      horizon slices, 61, 62
   time-lapse seismic data, 60–63
   water drive, 62
biodegradation of oil, 81
borehole seismic data, 86–87

crosswell geometry, 87
crosswell tomography, 87
   vertical seismic profile, 86
boundstones, 93, 94
bright spots, 60, 91
bubble point, 57
Buttonbed Black Shale, Coalinga field, California, 85
bypassing of oil, 54, 59, 60, 61, 62, 103, 115, 117

**C**
calcretes, 97
California, 84–85
Canada, 45, 81, 99, 101, 103–105
capital outlay (*Co*), 10
carbonate platforms (*see* platforms, carbonate)
carbonate-reservoir parameters, 98–100, 102
   clastics, compared with, 98–99
   Gassmann theory, 98–100
   pore aspect ratio, 98–100, 102
   pore shape and, 98
   prediction of, 98
carbonate rocks, reservoir geophysics of, 91–105
   clastic rocks, comparison with, 91
   correlation lengths, 92, 93
   diagenesis, effects of, 95, 96
   epeiric, 95
   facies and prediction of facies, 93, 94
      reservoir-prone facies, 94
      seal-prone facies, 94
      source-prone facies, 94
   fractures, 100–105
   Ghawar field, Saudi Arabia, 97–98
   lithic characteristics, general, 91–92
   Malampaya field, 95, 96, 97
   platforms, 91, 92
   reservoir characterization, 96–98
   reservoir heterogeneity, 92
   reservoir monitoring, 99–100, 101, 102, 103–105
      in fractured carbonate rock, 103–105
   seismic facies mapping, 92–95
   special considerations, 98–99
   systems tracts, 93
   volumetric characterization, 96
carbonate rocks, reservoir monitoring in, 99–100, 101, 102, 103–105
channel, abandonment, 59
channel, incised, 49
channel-fill sandstones, anatomy, seismic response, 17, 27, 44, 66
   and high-amplitude response, 44
channeling, 4, 54, 118, 119
channels, fault-controlled, 46
channel system, 59
clastic reservoirs, 41
Coalinga field, California, 84–85
   Buttonbed Black Shale, 85
   Kreyenhagen Shale, 85
   Temblor Formation, 85
Cobra problem, 115–119 (*see also* Cobra project)
   amplitude map, 116, 117, 118
   bypassed oil, 115, 117
   channeling, of water, 118, 119
   difference volume, 118
   fault block, 117
   4D seismic view, 115–116, 118
   gas cap, 116
   horizon slices, 116, 117, 118, 119
   results, 115
   sweep, uneven, 117
   3D seismic survey, 115, 118, 119
   time-lapse effects, 115, 119
   time line of activity, 116
   2D seismic survey, 118
Cobra project, 65–79 (*see also* Cobra problem)
   amplitude map, 77
   compartments in, 74, 77

cross section, relative impedance, 76
cross section, seismic, 75
delineation wells, 69
depositional environments of sands, 66, 67
development wells, 67, 69, 70, 78
   placement of, 78
fault blocks, 65, 66, 68, 75, 76
fluids, map of, 68, 69, 70, 71, 72, 73, 77
gas-oil contact, 67
geology, 65–66
history of development, 66–73
horizon slice, 74, 77
hydrocarbon anomalies, 75
investment, opportunity for, 79
oil-water contact, 67, 69
prediction of porosity and fluids, 74
primary production, decline of, 69
production, over time, graph of, 73
production history, 69–71
reservoirs, stacked, 74
secondary recovery, 71–73
sequence stratigraphy, 66
structural geology, 66, 68, 69, 75, 76
3D seismic survey, 74–77
time line, 75
2D data, 65
type log, 68
water, movement in reservoir, 70–73
coherency slice, 48
   carbonate-rock reservoir, 48
compaction of reservoirs, 53, 57, 58
compartments and compartmentalization, 4, 5, 18, 19, 33, 41, 50, 52, 54, 58, 74, 77, 83, 85, 109, 111, 113
   drainage and, 52
   faults and, 54
   flow models and, 52
cone-and-cylinder image, of VSP, 88
coning of water, 54
connectivity of reservoir, 52, 84
   effects of, in gas reservoirs, 84
   effects of, in heavy-oil reservoirs, 84
   effects of, in light-oil reservoirs, 84
   geochemical analysis and, 52
   well-flow tests and, 52
coralgal boundstones, 93
coralgal floatstones, 93
correlation, seismic data to well data, 16–17
$CO_2$ injection, 58, 103–105
   Weyburn field, Canada, 103–105
Cretaceous sands, Athabasca, Canada, 82
critical porosity, described, 45
cross-equalization, 116
crosswell electromagnetics, 63, 88, 89
   and bypassed oil, 88
   and clay, 88
   and porosity, 88
   and water saturation, 88
crosswell seismic data, 63, 89
crosswell tomogram, 90
   of velocity, compared with electromagnetics, 90

**D**
decision tree, 9
delineation, case history, 22–27
delineation and appraisal, contrasted, 13
delineation of oil field, 29–40 (*see also* Salinas project)
   decision table, 35
   geophysical program, options, 36
   investment profile, 34
   key information, 29–33
   objective, 29
   reservoir engineering, information, 34
   seismic investment opportunity, 35
delineation of reservoirs, 20–22
   factors in, 20
   strategy for, 20–22
   type of reservoir, seismic data suited for, 22

Distinguished Instructor Short Course • 125

delineation phase, of field development, 4, 5
delineation programs, purposes, 21
delineation-well program, 20, 21
    factors in, 20
    purposes, 21
delineation wells, purpose of, 33
delta-front sands, 66, 67
deltaic sands, 66
density, correlated with impedance and velocity, 75
density logs, 14, 16, 68, 74, 75
depositional cycles, 49
depth uncertainty, 18
deterministic models, 8
development drilling, 42, 44
development of reservoirs, 41–52
    attributes workflow, 47
    economic drivers, 41
    inversion, 48
        case example, 49–50
    optimized, 41
    reservoir characterization, 41–43
    reservoir flow model, 50
    seismic attributes, 44–47
    seismic response, 43–44
    strategy, 51–52
development phase, of field development, 4, 5, 52
    and investments, 52
development strategies, and subsurface geology, 41
development strategy, for production, 41
development strategy, reservoirs, 51–52
diagenesis, carbonate rocks, 95, 96
    and seismic-reflection continuity, 95
    and sequence stratigraphy, 95
    common effects, 96
Diana appraisal wells, 19
    forward seismic model, 19
    reservoir architecture, 19
difference volume, 118
digital recording, 3
direct arrivals, 17
direct hydrocarbon anomalies, 75
directional drilling, 4
dissolution pores, 44
downhole geophysics, 86
downlap, 93
drive mechanisms (see reservoirs, drive mechanisms)
drivers, economic, 41
drowned platforms, 91, 92
Duri field, Indonesia, 85–86, 87
    energy-amplitude maps, 87
    4D seismic map, 87
    intervention opportunities, 87
    pressure waves, 85–87
        seismic detection of, 87
    steam-flood monitoring, 85–86, 87

## E

economic drivers, of production, 53
electromagnetics, 2
energy-amplitude maps, Duri field, Indonesia, 87
energy envelope, 46
Eocene, 85
epeiric platforms, 91, 92, 95
evaporates, 92, 94
    platform, 94
expansion of nonreservoir rock, 58
exploration, 5
exploratory well, geology and seismic data, 13–15

## F

facies, and prediction of facies, carbonate rock, 93, 94
facies, and waveform shape, 45
facies, carbonate-rock, 93 (see also platforms, carbonate)
    prediction and limitations, 93
facies mapping, 92–95
facilities, installation of, 42

expenses, 42
fault blocks, 65, 66, 68, 75, 76, 117
faults, 2, 6, 43, 45, 46, 61, 62
fiberglass casing, 89
field delineation, 13–27
field management, 41
    development phase, purpose, 41
    factors in, 41
fingering of gas, water, 54
fishbone horizontal wells, 83
flooding surfaces, 67
flow-simulation model, 111
flow tests, 30
fluid dynamics, reservoirs, 53
fluid modeling of seismic data, 18
fluid-simulation modeling, 58
formation microimager, 30
formation volume factor (Bo), 8
forward models of seismic data, 17, 19, 47, 48, 56, 58
forward seismic-trace modeling, 47, 48
4D seismic data, 4, 55–63, 85, 87, 104–105
    (see also time-lapse seismic monitoring)
    and reservoir management, 62–63
    and heavy oil, 85
    map, Duri field, Indonesia, 87
    Weyburn field, Canada, 104–105
4D seismic view, 115–116, 118
fractures, and reservoir geophysics, 100–105
    amplitude and, 101
    and injection-well pattern, 103–105
    and shear-wave splitting, 101
    anisotropy and, 101
    azimuthal anisotropy and, 101, 102, 103
    orientations, relative to P-waves and S-waves, 102
fracturing, as reservoir stimulation, 4
frequency attributes, 46
frequency decomposition of seismic traces, 45

## G

gamma-ray logs, 14, 67, 74, 75
gas cap, 18, 116
gas caps, detection of, by AVO, 22, 24–25
gas injection, and amplitude difference, 59
Gassmann theory, 98–100
geologic models, 6
geophysics, defined, described, 1, 2–3
geostatistical models and reservoir characterization, 42
Ghawar field, Saudi Arabia, 97–98
    Kuff-C reservoir, 97–98
Girasol, Angola, and 4D seismic data, 58–59
    abandonment channel, 59
    bypassed oil, 59
    channel system, 59
    4D difference data, results of, 59
    gas injection, and amplitude difference, 59
    3D seismic data, 58, 59
        horizon slice, 59
    time-lapse seismic data, 58–59
    turbidites, 58, 59
grainstones, 93
grain-supported pores, 44
gravity, 2, 63
gross rock volume (GRV), 11
ground truth, 16
GRV (see gross rock volume)
Gulf of Guinea, 58
Gulf of Mexico, 60–63

## H

Habshan Formation, 95
Hamaca field, Venezuela, 84
heavy oil, 81–90
    borehole seismic data, 86
    business, 81–82
    Coalinga field, California, 84–85
    crosswell electromagnetics, 88
    crosswell seismic data, 89
    difference from light-oil production, 81
    Duri, Indonesia, example, 85–87

    4D seismic data, 85
    geophysics, application of, 84–85
        reservoir characterization, 84
        reservoir compartmentalization, 85
        reservoir connectivity, 84
        reservoir monitoring, 84
    horizontal wells and, 83
    problems inherent in, 82–83
    secondary recovery, 82–84
    subsurface geology, impact of, 83
    value of information (VOI), 89, 90
heavy-oil reservoirs, 54, 58, 88
    geophysical applications for, 88
horizon slice, 22, 24, 26, 27, 29, 31, 43, 44, 46, 48, 52, 55, 59, 61, 62, 74, 77, 103, 109, 111, 116, 117, 118, 119
    and depositional facies, 43
    and saturation of oil and gas, 55
    Bay Marchand field, Gulf of Mexico, 61, 62
    carbonate-rock reservoir, 48
    Cobra project, 74, 77
    fluvial sand compartments, 52
    inverted acoustic impedance, 52
    relation to facies, 26
    relation to fluids, 26
    Weyburn field, Canada, 103
horizon slice, and porosity, 7
horizon slice, demonstrated, 7
horizontal drilling, 4
horizontal wells, 83, 103
    and heavy-oil production, 83
    and steam-assisted gravity drainage, 83
huff-and-puff steam flooding, 86, 87
hydrocarbons, direct detection of, 3
hydrocarbons, origin of, 2

## I

impedance, correlated with density and velocity, 75
impedance attribute, 51
impedance log, 75, 112
incised channels, 49
Indonesia, 85–87
inversion, of seismic data, 14, 47–50, 74
    advantages of, 49
    and acoustic well logs, 47, 49
    case example of, 48, 49–50
    described, 47, 49
    for prediction of sands, 74
    limitations of, 49
isolated platforms, 91, 92, 93

## K

karst breccia, 94
Kern River field, California, 81, 82
Kreyenhagen Shale, Coalinga field, California, 85
Kuff-C reservoir, Ghawar field, Saudi Arabia, 97–98

## L

Lekhwair Formation, 95
"line-drive" injection, Weyburn field, Canada, 103

## M

magnetics, 2
Malampaya field, 95, 96, 97
    diagenesis, effects of, 95, 97
        calcretes, 97
        paleosol, 97
        sequence of, 97
        speleothems, 97
        vugs, 97
    Pagasa clastics, 97
    reflectors, 97
    seismic facies, 96
    time lines, 96
management of oil and gas field, 5
marine acquisition, 2
marl, as reservoir rock, 103–105
microresistivity log, 30
migrated gradient, 2D compared with 3D, 26
migration of seismic data, 26

# Index

importance, 26
2D compared with 3D, 26
migration stack, 2D compared with 3D, 26
miliolids, 93
Mississippian rocks, Weyburn field, Canada, 103–105
modeling, forward seismic, 17, 19, 47, 48, 56
modeling seismic responses, 16
models, matched with seismic lines, 44
mudstones, 94
mudstones-wackestones, 94
multiple noise, 109
multiples, 16, 109, 110
    subsalt multiples, 109, 110
Mut Basin, Turkey, 93
    Miocene platform, facies, 93

## N

net cash flow (*Ct*), 10
net percent of reservoir-quality rock (ntg), 8
net present value (NPV), 9–10, 13
neural networks, 45
neutron logs, 68
normal incidence reflection, 22
North Reef, 48
NPV (*see* net present value)
nummulitids, 93

## O

offshore fields and 3D seismic data, 91
oil bright spot, 60
oil rim, 18
oil-water contact (OWC), 8, 67, 69
onlap, 93
operational phases, information needed, 4
operations, oil and gas, 3–5
optimized development of reservoirs, 41
original oil in place (OOIP), 8, 9, 11, 21
    equation, 8
    parameters in, 8
    tornado chart, 11, 21
Orinoco belt, Venezuela, 84
overbank sands, 27, 31, 66
OWC (*see* oil-water contact)

## P

packstones, 93
Pagasa clastics, Malampaya field, 97
paleosol, 97
passive seismic recording, 2, 63
pinch-outs, 43
pitchfork horizontal wells, 83
platforms, carbonate, 91–95
    classification, 92
    drowned, 91, 92
    epeiric, 91, 92, 95
    isolated, 91, 92, 93
    ramps, 91, 92
    reservoir-prone facies, 94
    rimmed shelves, 91, 92
    seal-prone facies, 94
    shelves, 91, 92
    source-prone facies, 94
Pliocene, 85
point bar, 46
pore aspect ratio, 98–100, 102
porosity, mapping, seismic-reflection amplitude, 98
porosity reservoir-model attribute, 52
precipitation of solids and occlusion of porosity, 83, 84
prediction of reservoirs (*see* seismic attributes)
pressure maintenance, 62
pressure waves, 85–87
    seismic detection of, 87
production, 53–63
    economic drivers, 53
    geophysics and, 53–55
    time-lapse seismic data, 55–63
        Girasol, Angola, 58–59
        Bay Marchand, Gulf of Mexico, 60–63
production and geophysics, 53–55
    seismic reflectivity, 53

production rates, initial, 41
production stage, of field development, 4, 5
project value, measures of, 9–10
P-S mode conversion, 103
P-waves, velocity of, relative to fractures, 102

## R

Rayda Formation, 95
recovery, efficiency of, 42
    relation to geologic environments, 42
reflection continuity, and structural geology, 6
reflection seismology, 2
reflection strength, 6, 45, 46
    defined, 46
    rock and fluid properties, estimation, 6
    total, 45
    variation of, causes, 6
refraction seismic, 2
relative-impedance cross section, 76
relative permeability, 82
reserves, defined, 14
    possible, 14
    probable, 14
    proved, 14
reserves, verification of, 13
reservoir, defined, 1
reservoir attributes, 43
    and seismic modeling, 43
reservoir characterization, 41, 42, 84, 85, 96–98
    and geostatistical models, 42
    and seismic data, 42
    heavy-oil fields, 85
    of carbonate rock, 96–98
reservoir compaction, 53, 57, 58
reservoir flow model, 50
reservoir forecasting, 50
reservoir geophysics, 1, 91, 91–105
    carbonate rocks and clastic rocks, compared, 91
    defined, 1
reservoir management, 1, 53, 62–63, 65, 84
    and 4D seismic data, 62–63
    and sequence stratigraphy, 84
    requirements, 53
reservoir models, 20, 21, 53
    end-member, 20
    midrange, 20
    simulation models, 53
    static, 20
    uncertainty, measurement of, 21
reservoir monitoring, 53, 84, 87, 99–100, 101, 102, 103–105
    and seismic reflectivity, 53
    and time-lapse seismic data, 87
    in carbonate rocks, 99–100, 101, 102, 103–105
    in fractured carbonate rock, 103–105
reservoir parameters, 18, 98–100, 102
    carbonate rock, 98–100, 102
reservoir prediction model, 51
reservoir properties, prediction of, 5–6
    dynamic properties, 5–6
    static properties, 5
reservoir properties, seismic data, and well data, 13–14
reservoir properties and seismic data, limitations, 50
reservoir-quality rock, in trap, 8
reservoirs, delineation of, 20–22
    factors in, 20
    strategy for, 20–22
reservoirs, development of, 41–52
    attributes workflow, 47
    economic drivers, 41
    inversion, 47, 49–50
        case example, 49–50
    optimized, 41
    reservoir characterization, 41–43
    reservoir flow model, 50
    seismic attributes, 44–47

    seismic response, 43–44
    strategy, 51–52
reservoirs, drive mechanisms, 41
    bottom-water, 41
    edge-water, 41
    primary-pressure, 41
    reservoir compaction, 41
reservoirs, optimized development of, 41
reservoirs, prediction of, and seismic attributes (*see* seismic attributes)
reservoirs, seismic models of, 41
resistivity log, 68
resistivity of the formation, and steam flooding, 88
rimmed shelves, 91, 92
rock and fluid properties, estimation, 6
rock velocity, 47

## S

sag, mapped, 102
SAGD (*see* steam-assisted gravity drainage)
Salil Formation, 95
Salinas problem, 107–113 (*see also* Salinas project)
    abandonment channel, 109
    amplitude attribute, 109
    amplitude map, 107
    cash flow, 113
    compartmentalization, 109, 111, 113
    delineation, results of, 107–108
    development, 108–111
    flow-simulation model, 111
    gas-water contact, 108
    horizon slice, 109, 111
    prestack and poststack depth-migrated data, 110
    reservoir model, 111, 113
    salt, 107, 109, 110
    salt, effects of, 109, 110
    seismic lines, 108
    seismic sections, 110, 112
    simulator model, 112
Salinas project, 29–40 (*see also* Salinas problem)
    amplitude anomaly, 29, 31, 32
    amplitude map, 31, 36
        correlated with reservoir quality, 31
        poststack-migrated, 36
    conceptual model, 31, 32
    decision table, 35
    discovery well, reservoir parameters, 33
    horizon slice, 31
    investment profile, 34
    overbank deposits, 31
    reservoir sandstone, log, 30, 32
        parameters, discovery well, 33
        porosity, relative to amplitude, 32
    salt-dome overhang, 29
    sand-channel system, 29, 31, 32
    seismic cross section, 31
    seismic investment opportunity, 35
    seismic lines, 36, 39, 40
    seismic trace, synthetic, and gas amplitude anomaly, 32
    seismic traces, 36–38
    spill point, 29
    stratigraphic model, 32
    structural geology, 30
    subsalt imaging, 29
salt, 8, 15, 107, 109, 111
salt dome, 29
salt-dome overhang, 29
sandbars, 51
sand-channel system, 29, 31, 32
sandstone depositional environments, and recovery efficiency, 15
sandstones, and barriers to flow, 15
saturation images of reservoir, 88
Saudi Arabia, 97, 98
seabed multiples, 17
seals, of reservoir, 2
secondary recovery, 5, 41, 71–73, 82–84
    Cobra project, 71–73
    heavy oil, 82–84

seismic amplitude, 44
seismic attributes, 13, 41, 44–47, 51, 74
    amplitude, 45, 46, 51
    as proxy for geology, 13
    AVO fluid factor, 51
    coherence, 45, 46
    for prediction of fluids, 74
    for prediction of reservoirs, 44–47
    frequency, 45, 46
    impedance, 51
    period, 46
    reflection strength, 46
    shape, 45, 46
    velocity, 45, 46
seismic attributes, seismic data, 3D, 41
seismic classification and stratigraphic interpretation, 47
seismic-classification map, 47
seismic cross section, 6, 15, 31, 75
seismic data, trace form recast in amplitude form, 45
seismic data and reservoir models, 6–7
seismic-data attributes, 48
    workflow, 48
seismic facies mapping, carbonate-rock, 92–95
seismic lines, 36, 39, 40
seismic modeling, 43
    and reservoir attributes, 43
seismic models of reservoirs, 41
seismic patterns, 43
    and faults, 43
    and pinch-outs, 43
    and stratal patterns, 43
    and unconformities, 43
seismic reflection, 3
seismic-reflection amplitude, 98
    and porosity, mapping of, 98
seismic reflectivity and reservoir monitoring, 53
seismic refraction, 3
seismic response, 15, 19, 43, 44
    and diagenesis, 43
    and flow properties, 15
    and fractures, 43
    and grain sorting, 43
    and grain-to-grain contacts, 43
    and lithology, 43
    and pore shape, 43
    and pressure, 43
    and water saturation, 19
    variables in, 44
seismic response models, 19
seismic time difference, mapped, 102
seismic tomography, 88, 89
seismic traces, 36–38, 45, 47
    and 3D coherence, 45
    classification, 45, 47
    frequency decomposition of, 45
    horizon slice, 45
    spectral decomposition of, 45
sequence stratigraphy, 43, 66, 84, 93, 95
    and reservoir management, 84
shadow zone, 8
shale and shales, 2, 65, 67
    seals, 65, 67
    sources, 65
shape attributes, 47, 48
    and waveform character, 47
    workflow for selection of, 48
shear-wave splitting, 101
    and fractures, 101
shelf sands, 65
shingled channel axes, 17
shoals, 93
sidewall cores, 30
sidewall samples, 14
slumps, 93
smearing, 24
sonic logs, 16, 50, 75
South Reef, 48
spectral decomposition of seismic traces, 45
speleothems, 97

spill point, 29
statistical models, and prediction of reservoir properties, 8
steam-assisted gravity drainage (SAGD), 82, 83
    horizontal wells and, 83
steam-flood monitoring, 85–86, 87
    and forward seismic modeling, 85–86
    and pressure waves, 85
    Duri field, Indonesia, 85–86, 87
    forward seismic modeling, 85–86
steam flooding, 41, 58, 82, 83, 85–87, 88
    and formation resistivity, 88
    choke installation, 86, 87
    Duri, Indonesia, steam-flood monitoring, 85–86, 87
    forward seismic modeling, 85–86
    huff and puff, 86, 87
    injector shut-in, 86, 87
    steam isolation, 86, 87
    time-lapse seismic data and, 86, 87
strategic fit, 9–10
stratigraphic model, 18, 32
stratigraphy, and recovery of oil, 15
streamers, 2
structure map, 7, 24
subsalt imaging, 29
subsurface, information, relative value of, 10–11
subsurface geology, 83
    and heavy-oil production, 83
S-waves, velocity of, relative to fractures, 102
sweep management, 62
synthetic inverted traces, 50
synthetic seismic models, 16
synthetic seismic traces, 16, 32, 74, 75
    and gas amplitude anomaly, 32
synthetic seismograms, 44, 45
    1D, 44, 45
    and seismic attributes, 44–45
systems tracts, carbonate, 93
T
talus, 94
tar mat, 81
Temblor formation, Coalinga field, California, 85
tertiary recovery, 103–105
Tertiary section, 65
thief zones, 83
3D coherence, 45
    and faults, 45
    and seismic traces, 45
3D prestack depth migration (PSDM), 34
3D seismic data, 4, 24, 25, 26, 91
    collapse of image, 24
        compared with 2D seismic, 24, 25, 26
    offshore fields and, 91
3D seismic surveys, 60, 61, 62, 74–77, 103, 104, 115, 118, 119
    Bay Marchand, 60, 61, 62
    Cobra problem, 115, 118, 119
    Cobra project, 74–77
    Weyburn field, Canada, 103, 104
3D stratigraphy, 3
tidal channels, 92
tidal-flat sands, 67
time advances of reflections, 57
time delay of reflections, 57
time-depth conversion, 18
time-lapse effects of production, 115
time-lapse seismic data, 86–87
    and reservoir monitoring, 87
    and steam flooding, 86
time-lapse seismic monitoring, 55–63, 104–105
    attributes, differences, and production, 57
    Bay Marchand, Gulf of Mexico, 60–63
    evaluation of potential, 57
    evaluation of results, 57–58
    Girasol field, Angola, 58–59

    objective, 57
    results of, examples, general, 58
    Weyburn field, Canada, 104–105
    workflow, 56
time-lapse surveys, 3
    and fluid movement, 3
    and pressure, 3
tomogram, crosswell, 90
    velocity, compared with electromagnetics, 90
tomography, 87, 88, 89, 90
tornado chart, 11, 21
total reflection strength, 45
"trillion-dollar tar pit," 82
turbidites, 23, 58, 59, 94
turbidity currents, 25, 26
    channel deposits, 25, 26
        AVO response, 26
    overbank, deposits, 25, 26
Turkey, 93
2D seismic data, limitations, 65
U
uncertainties, and reservoir models, 8
uncertainty, illustration of, 7–8
uncertainty, limits on, in reservoir models, 50
uncertainty about reserves, measurement of, 21
uncertainty tools, use of, 7–8
    illustration of, 7–8
unconformities, 2, 6, 43, 93
    convergence of, 93
V
valuation of resource, elements in, 10
value of information (VOI), 8, 9, 20, 56, 57, 89, 90
    and decision trees, 9
    procedure, 57
velocity, correlated with density and impedance, 75
velocity, direct measurement of, 89
velocity, of seismic data, 45
Venezuela, 50, 81, 84, 103
vertical seismic profiling (VSP), 16, 17, 86–87, 88
    resolution compared with surface seismic data, 86–87
    3D images, for reservoir characterization, 87
    3D images, for reservoir monitoring, 87
visualization and seismic data, 7
VOI (see value of information)
volume-of-shale log, 68
volume seismic attributes, 45, 48
    coherency attribute, 48
    minimization of, 45
        neural network prediction, 45
volumetric characterization, carbonate-rock reservoirs, 96
$V_P/V_S$ crossplot, 99
VSP (see vertical seismic profiling)
vugs, 97, 102
W
water breakthrough, 70, 71–73, 83
water flooding, 41, 83
water saturation, 19, 60, 88
    and crosswell electromagnetics, 88
    and diminishment of seismic response, 19, 60
waveform shape and facies, 45
well data, extrapolation, 18–19
western Canadian basin, 82
Weyburn field, Canada, 103–105
    anhydrite seal, 105
    bypassed oil, 103
    $CO_2$ injection, 103–105
    4D seismic monitoring, 104–105
    fractures, and injection-well pattern, 103–105
    "line-drive" injection, 103
    3D seismic surveys, 103, 104